T0408679

Creative Chemists

Strategies for Teaching and Learning

Advances in Chemistry Education Series

Titles in the Series:
1: Professional Development of Chemistry Teachers: Theory and Practice
2: Argumentation in Chemistry Education: Research, Policy and Practice
3: The Nature of the Chemical Concept: Re-constructing Chemical Knowledge in Teaching and Learning
4: Creative Chemists: Strategies for Teaching and Learning

How to obtain future titles on publication:
A standing order plan is available for this series. A standing order will bring delivery of each new volume immediately on publication.

For further information please contact:
Book Sales Department, Royal Society of Chemistry, Thomas Graham House, Science Park, Milton Road, Cambridge, CB4 0WF, UK
Telephone: +44 (0)1223 420066, Fax: +44 (0)1223 420247,
Email: booksales@rsc.org
Visit our website at www.rsc.org/books

Creative Chemists
Strategies for Teaching and Learning

Simon Rees
University of Durham, UK
Email: simon.rees@durham.ac.uk

and

Douglas Newton
University of Durham, UK
Email: d.p.newton@durham.ac.uk

Advances in Chemistry Education Series No. 4

Print ISBN: 978-1-78801-511-0
PDF ISBN: 978-1-83916-109-4
EPUB ISBN: 978-1-83916-125-4
Print ISSN: 2056-9335
Electronic ISSN: 2056-9343

A catalogue record for this book is available from the British Library

The Royal Society of Chemistry is a charity, registered in England and Wales, Number 207890, and a company incorporated in England by Royal Charter (Registered No. RC000524), registered office: Burlington House, Piccadilly, London W1J 0BA, UK, Telephone: +44 (0) 20 7437 8656.

For further information see our web site at www.rsc.org

Printed in the United Kingdom by CPI Group (UK) Ltd, Croydon, CR0 4YY, UK

Foreword

I have spent most of my career teaching chemistry, and it has given me a lot of pleasure. When I think of the high points, they are mostly connected with moments when students show they are inspired by the subject that I am teaching them.

Most chemistry teachers have the same motivation – they measure their success in terms not only of examination results, but of students' being inspired and engaged by the subject and wanting to continue its study to the next level. Creative teachers are constantly thinking of new ways to engage students' imagination, helping them to understand more securely and inspiring them to want more.

Creative thinking is not just for teachers – it is for their students too. One of the many reasons why chemistry qualifications are so sought-after by employers is that they bring with them transferable skills: essential skills like teamwork, problem solving – and creativity. Employers value these skills because they are essential elements of human working behaviour that are hard to automate. Chemistry teachers can help students to become more employable by showing them how to think creatively.

This is a book about how to teach, and learn, chemistry creatively, to the benefit of both teacher and student. It is rooted in research evidence, but full of practical ideas and activities that teachers can use to engage and stimulate their students' own creative processes. All the essential elements of good chemistry teaching are here: chemical language, experimentation, performance, storytelling, assessment and more. There is no conflict between teaching creatively and teaching a knowledge-based curriculum, as these practical, evidence-based examples show.

Advances in Chemistry Education Series No. 4
Creative Chemists: Strategies for Teaching and Learning
By Simon Rees and Douglas Newton

Published by the Royal Society of Chemistry, www.rsc.org

One of the great fallacies about science is that it is dry, factual and inhuman – in apparent contrast to the arts and humanities. Nothing could be further from the truth: the practice of scientific research is a creative activity, and so should its teaching be. The authors' hero Michael Faraday showed this 150 years ago, and so it should be today.

Professor Sir John Holman
York

Preface

The motivation to produce this book originated from a student who had done very well in previous exams and had chosen biology, chemistry and psychology as their options to study from the age of 16–18 years old. The student had a clear interest in chemistry and started the new course with enthusiasm and engagement. However, within two weeks the student was complaining that none of their options enabled them to explore and express their creative side and they promptly dropped chemistry and opted to study drama instead. Many chemistry teachers can probably recount similar experiences.

How can we, as chemistry educators, enhance the creative aspects of the subject, provide opportunities for students to develop their creative abilities and gain an understanding of the importance of creativity in science? Arts subjects are often specifically referred to as the "creative arts". This distinguishes these subjects from the sciences which are therefore, by implication, not creative. This book is intended to stimulate conversations about creative pedagogies based on the published evidence and the experience of chemistry educators. It aims to deepen an understanding of creativity and how it applies to chemistry so that the phrase "creative sciences" is as widely accepted and understood as it is in the creative arts.

Producing this book led to a yearlong enjoyable and creative collaboration during which we identified ideas to engage a wide range of students in chemistry and thinking creatively. This book is intended for educators involved in teaching chemistry at any level of education. There are principles, theory and ideas that are applicable across the full range of education. Furthermore, this book is aimed at those chemistry educators who work with diverse students. We explore ways to engage more students with chemistry, recognise its relevance in their lives, and see the wide variety of skills that studying the subject can develop. It will also appeal to anyone involved in

Advances in Chemistry Education Series No. 4
Creative Chemists: Strategies for Teaching and Learning
By Simon Rees and Douglas Newton

Published by the Royal Society of Chemistry, www.rsc.org

science education or those interested in learning more about the links to their own subject areas. The book can be read throughout its entirety or can be dipped into for those areas of most interest.

In the first chapter, we establish the importance of developing creative thinking and imagination in this technological age. We explore the principles of creative thinking, its components and how it applies to both teachers and students in chemistry. This theme is discussed further in Chapter 2, where we discuss how the different aspects of creative thinking can be developed in the chemistry classroom. Divergent, convergent, associative and lateral thinking are explained and exemplified within the chemistry curriculum. Chapter 3 applies teaching creatively in the context of multisensory learning. A range of strategies and case studies are discussed that demonstrate how multisensory experiences can help students smell, taste and touch chemistry as well as visualise the sub-microscopic world. The importance of incorporating the human element into chemistry learning is explored in Chapter 4 in order to engage a broad range of students. Chemistry is culturally situated within a range of contexts inside and outside the classroom.

Chapter 5 specifically looks at how creative thinking is applied to construct and represent understandings in chemistry. Using the established design of the Periodic Table we illustrate how creative thinking can be used to identify shortcomings, leading to the development of new ideas which, in turn, can lead to new insights and perspectives. Multiple representations in chemistry are discussed, as are and how we can support students to develop mental models of the sub-microscopic world. Storytelling is central to our engagement in many aspects of our lives and, in Chapter 6, we explore the important principles of effective stories such as: unexpected change, curiosity, and appropriate detail. We exemplify the range of stories that can be told in chemistry and the variety of stimuli that can be used. This theme is continued in Chapter 7, where we consider the important aspects of performance and the use of drama based activities in chemistry. Using Michael Faraday as a case study, we identify the key principles of successful performance. We discuss examples of the application of drama and how to effectively organise scientific debate.

Practical chemistry is considered in Chapter 8, with discussion of the different types of practical investigation focusing on creative approaches, such as escape rooms and microscale chemistry. Central to creative thinking is the ability to use appropriate subject specific language to formulate models and ideas and to be able to discuss these with others. Chapter 9, therefore, discusses the specific language challenges that chemistry presents for students, and offers creative strategies to overcome these. A common challenge with incorporating creativity within formal education is the perceived need to assess it and how this can be achieved. These issues are explored in Chapter 10 with concrete examples of possible ways to assess creative thinking. Finally, Chapter 11 discusses why creativity matters.

This book is a celebration of the creative approaches within the chemistry community. It also supports the recommendations of the Durham Commission on Creativity and Education report.† Finally, we would like to thank many colleagues for their support and willingness to share their own innovative strategies and experiences. We would also like to thank Roy Alexander and Philip Stewart for comments and contributions to the section discussing the design of the Periodic Table and Vivian Mozul for comments on the Practical Chemistry chapter.

Simon Rees and Doug Newton

†https://www.dur.ac.uk/creativitycommission/report/

Contents

Advances in Chemistry Education Series No. 4
Creative Chemists: Strategies for Teaching and Learning
By Simon Rees and Douglas Newton

Published by the Royal Society of Chemistry, www.rsc.org

CHAPTER 1

Creative Teaching and Creative Students

1.1 Creative Thinking

Creative thinking in the classroom is seen as something to be promoted. Governments see it as a source of economic success, and it has personal value as it confers a certain autonomy on its holder through its potential to help with life's problems. Moreover, the digital revolution is making it possible to automate much that is routine in the workplace (Bakshi *et al.*, 2015). What will be left will be what takes imagination, heuristic thinking, and 'What if?' thinking. This is something for which we should prepare our students. Attention tends to focus only on students' creative competence (*e.g.* UNESCO, 2001; Shaheen, 2010), but it is also for the teacher (Rinkevich, 2011). Creative thinking, particularly in times of rapid change, makes teaching flexible so it can adapt to new needs and expectations. In a real sense, competence in creative teaching can make teachers future-proof as it helps them change with the times, even when robots appear in the classroom to take over routines. But it does more than that: creative teaching has probably always been worthwhile because it can produce teaching events which enhance learning. If the world and its students were always the same, then teaching might be a routine of recycled lessons. But no class is exactly like another, and expectations are always changing, so teaching is often better if it takes note of this. Creative thinking is as much for the teacher as for the student. Here, we aim to illustrate both aspects of creative thinking, that of the student and that of the teacher. But first, we need to set out what we mean by creative thinking.

There are many definitions of creative behaviour, but they tend to agree that its main component is the construction of something more or less new, novel or original (Acar *et al.*, 2017; Said-Meturaly *et al.*, 2017). Simply

Advances in Chemistry Education Series No. 4
Creative Chemists: Strategies for Teaching and Learning
By Simon Rees and Douglas Newton

Published by the Royal Society of Chemistry, www.rsc.org

producing something new, novel or original, however, is not enough otherwise any crazy idea would do. In addition, it also has to be somehow appropriate, fit for purpose, useful, or of value (NACCCE, 1999; Runco *et al.*, 2005). In addition, it helps if it is also somehow satisfying. Exactly what original, appropriate, and satisfying mean depends on the context.

1.2 Being Creative in Chemistry

The creative teaching of chemistry is clearly not the same as students thinking creatively about chemistry. Creative teaching is when a teacher applies imagination to produce novel approaches with the intention of making students' learning more interesting and effective. These approaches must be in some way novel to the teacher, but not necessarily novel to the world – it is conceivable that, a thousand miles away, another teacher is teaching in the same way. What is important is that the approach is appropriate in the sense that it has some promise of achieving its goal. If, at the same time, it is satisfyingly economical in terms of effort, time and resources, all the better.

Fostering students' creative thinking in chemistry aims to have students produce explanations or ideas that are novel to them, and, at the same time, appropriate in the sense that they are scientifically plausible. If those ideas are also concise or parsimonious, or otherwise cognitively economical, all the better (Rosch, 1999). Teaching which deliberately offers opportunities for students to practise their own creative thinking in chemistry is aimed at developing students' competence in it (Jeffrey and Craft, 2004; Rinkevich, 2011).

It is possible to have creative teaching without exercising students' creative thinking. For instance, a teacher may construct a novel way of teaching the arrangement of the Periodic Table after which the students are able to produce the table faultlessly from memory. Equally, there can be students' creative thought without creative teaching: a routine lesson about the Periodic Table may prompt students to reflect in a 'What if?' way about the chemical nature of non-carbon based life. Creative teaching and students' creative thinking, however, are both valuable aspects of chemistry education which offer benefits for both students and their teachers, and they can be mutually supportive. Creative teaching can be deliberately aimed at stimulating creative thinking, and, in the process, reap rewards which are greater than either alone. We begin with some thoughts about creative teaching and what it can do.

1.3 Creative Teaching in Chemistry

Creative teaching can produce rewards for the both teacher and student. Not least amongst these, creative teaching can:

- prevent teaching being seen by students as irrelevant, boring, or outmoded, and hence
  - attract and retain STEM students (Holbrook, 2005);

  - offer teaching which is tuned to the particular students in front of you in order to engage them effectively and, so, enhance their learning (Darby, 2005; Gibson, 2010).
- prevent teaching becoming a tedious treadmill of transmitted information, and hence
  - maintain teachers' own interest in their subject and its teaching (Craft *et al.*, 2014);
  - enhance job satisfaction;
  - maintain the teachers' enthusiasm, and through emotional contagion, enhance the emotional climate of the classroom and engagement in learning (Newton, 2014);
  - enable a teacher to adapt effectively to changing needs, expectations, and subject content, and, in effect, possess a competence which future-proofs them.

How are these rewards to be achieved? One way is to see the subject from the *student's perspective*, in particular, to ask: What is in it for them? (Craft *et al.*, 2014). The temptation can be to point to the practical utility of the topic, but this can be self-defeating as some knowledge has little obvious or immediate utility but opens the way to later learning. Instead, the question to ask is: What are the students' psychological needs that studying this topic will satisfy? For instance, these might be the satisfaction of curiosity, a need to feel competent, a need to understand the world and one's place in it, and a need to affiliate with others. Relating the topic to the students' needs is what can make it relevant in the eyes of the student. Science has tended to purge its teaching of people, possibly to reflect or emphasise the objective nature of the subject, or, perhaps, those who teach it (Newton, 1988). But, putting people, individually and collectively, back into what is taught can help students relate to and see the relevance of a topic. This is where imaginative, creative thought helps a teacher literally bring a topic to life, and do so without selling short the topic in hand. This might be achieved in a variety of ways. For instance, some have had success using drama, ontological questioning and collaborative activities (Pollard *et al.*, 2018).

Seeing the subject through the students' eyes also means deliberately planning to support emotional needs, like interest. A teacher's enthusiasm can attract interest and attention as students look for the source of enthusiasm. (Of course, excessive enthusiasm can be a source of amusement, rather than an attraction, and interest will die away if relevance does not become evident.) Pollard *et al.*, (2018) have pointed to the engaging nature of surprise. This does not mean that the surprise must always come from what has been called Whizz-Bang chemistry – students may remember the whizzes and bangs but not the chemistry – but, for example, it can be more muted and felt as the pleasure at finding a novel way of solving a problem. Needing to feel secure, and feeling able to offer an idea or make mistakes without fear of ridicule, are also states which support cognitive engagement in learning (Darby, 2005).

Such approaches can catch students' interest, engage them in thinking, make learning more memorable and durable, motivate students to want more, and enhance their attainment: all are valuable outcomes of the teacher's creative efforts. These efforts may be usefully directed at any kind of purposeful thinking in chemistry. In this context, purposeful thinking is aimed at achieving particular academic ends, often to develop knowledge, know-how, and thinking habits, and ranges from memorising information to higher level mental activity, like evaluative or critical thinking. (There is a tendency to deride so-called 'lower' level thinking, like memorising, but all domains contain information which may usefully be committed to memory; the problem comes when students try to commit everything to memory – as when cramming for examinations – when we want them to develop competence in other kinds of purposeful thought.) Kinds of purposeful thinking which creative teaching could support include:

- memorising
  - *as when learning the symbol used to indicate a poisonous substance, or that there is a limit to the amount of a substance which will dissolve in water, or that chlorine is a green, diatomic gas which can purify water, and mercury is a metallic element of atomic number 80 and symbol Hg;*
- deducing
  - *as when a chemical equation is made to balance, a molecular formula is deduced in analytical chemistry, reaction pathways are inferred in complex chemical systems, or when the soil's pH is deduced from the colour of certain plant foliage;*
- understanding
  - *as when grasping the notion of permanent and temporary changes, or chemical bonding, or knowing why chlorine can purify water, or why buckminsterfullerene has properties not usual amongst other materials, or how the elements in the groups of the Periodic Table relate to one another, or constructing an analogy;*
- creative thinking
  - *as when Dalton proposed that, ultimately, matter comprised indivisible, discrete particles, and that those of the same element are identical. More recently, as when the concept of lock and key was constructed to explain biomolecular recognition, and the as yet unsolved problem of constructing a material which usefully superconducts electricity at room temperature;*
- evaluative and critical thinking
  - *as when evaluating the plausibility of a claim or belief such as, if wine is left to evaporate, the alcohol in it will become more concentrated, or considering the adequacy of an experiment to test a tentative explanation, or the strength of the evidence for a conclusion.*

We are expected to help students of all ages exercise these kinds of purposeful thinking (Newton and Newton, 2018). Economy of effort, the shadow of examinations, and time pressures may incline some students and their teachers to focus on some of these more than others. One that can be overlooked is students' creative thinking. This is not so say that other kinds are not important: there can be no creative thinking without them (Newton, 2014; Hirsch, 2017).

Of course, creative teaching need not end here. Chemistry teachers put creative effort into constructing new programmes and new approaches, and these sometimes fit the current and anticipated needs of science education so well that they may be the basis of an invigorating reformulation of science teaching. As with many such ideas, the constraints of the classroom and prescribed schemes of work can make the creation easier than its adoption, but it can and does happen from time to time.

1.4 Students Thinking Creatively in Chemistry

In science, highly regarded creative thinking is generally seen as being:

- original;
- plausible or appropriate, according to context;
- parsimonious or cognitively economical.

Of these, the first two are generally essential and the last is desirable or welcome (Newton, 2016). Popularly, there is a tendency to attach the word 'creative' to the arts as though the arts have the monopoly on creative activity. Of course, there is creative thinking in domains like the sciences and mathematics, although it may appear under the guise of non-algorithmic problem solving (Claxton, 2006).

Creative thinking, and its twin, problem solving, can be seen as constructing personal understandings (Newton, 2012). In the classroom, the construction of an understanding often means bringing together new information, making connections within it, and relating it to prior knowledge to give a functional model of some aspect of the world. This mental model, however, is generally what is approved by others (the teacher, the textbook, and chemists collectively). When creative thinking brings together seemingly unrelated ideas, the understanding that results may be novel. The writer, Arthur Koestler, described this process as bi-sociation (Koestler, 2017). It is facilitated by the imagination, a mental activity which allows play with seemingly disparate ideas in a 'What if?' way to produce alternative worlds. In a sense, understanding lies at one end of a spectrum while at the other is more adventurous, self-determined creative thinking.

Dijrksterhuis and Nordgren (2006) have provided a useful, theoretical framework which has relevance for creative thinking. They describe two modes of thought, the conscious and the unconscious. Conscious thought is deliberate, controlled, attention-dependent, low-capacity, relatively slow, sequential, analytical, and rule-following. Unconscious thought, however, is automatic, involuntary, high-capacity, fast, and associative. Different kinds of purposeful thought may draw on these two modes to different degrees. Careful, step-by-step deduction, not uncommon in chemistry, mathematics, and other sciences, generally requires focused attention and rule observance. Constructing an understanding of a chemical structure, on the other hand, is likely to benefit from some imagination which brings together ideas. Dijrksterhuis and Nordgren add that, 'conscious thought stays firmly under the searchlight, [whereas] unconscious thought ventures out into the dark and dusty crannies of the mind' (p. 102). On this basis, the potential for making associations in unconscious thought can be a useful resource in creative thinking.

Fostering students' thinking in chemistry is worthwhile for several reasons. First, creative thinking can be emotionally rewarding and highly motivating. By necessity, opportunities to be creative offer students some mental freedom to explore, follow a thought, test ideas, and even make mistakes without unfavourable judgements being made. This kind of autonomy supports self-determination, fosters the ability to control events and cope in a rapidly changing world, and allows students to face an uncertain future with a feeling of self-efficacy and well-being (Black and Deci, 2000; Lewis *et al.*, 2009; Shaheen, 2010; Taannaeifer and Motaghedifard, 2014).

Second, having some competence in thinking creatively can support the ability to solve ambiguous problems, and is potentially useful in the job market (Newton, 2012). Recent reports point out that the digital revolution brings with it a change in the nature of work. In particular, work which can be captured by routine procedures or algorithms may be automated or delegated to robots. In the very near future, this would put at risk at least one-third of jobs in the UK and USA. On the other hand, occupations which cannot be described by routines are more likely to survive or increase (Bakshi *et al.*, 2015). Those who can work effectively in non-routine ways, like thinking creatively, may be at a premium. At the same time, their products can support a country's economy, something which is of interest to governments worldwide (Shaheen, 2010).

Third, Holbrook (2005) has pointed to the need for chemistry to be perceived as relevant if students are to value it, find it attractive, and consider it a worthwhile career option. To that end, he urged that more attention should be given to cognitive skills which go beyond accumulating information and procedural algorithms. Indeed, Ball (2006) has highlighted the need for chemistry to be seen as worthwhile in its own right, rather than only as a tool to support other sciences.

Taken together, this indicates that some competence in creative thinking in chemistry could foster students' interest, enhancing their motivation, be satisfying and fulfilling, and, potentially, add to their life chances. With some prescience, Piaget noted a need to foster creative thinking in the classroom. His view was that, 'The principal goal of education in schools should be creating men and women who are capable of doing new things, not simply repeating what other generations have done' (see Newton, 2017). But what does creative thinking in chemistry mean?

There are several kinds of creative thinking students could practise in chemistry, mirroring those experienced by the practising chemist. We centre these upon the following creative spaces.

- ***The problem space***
  - Many years ago, Torrance (1999) and Guilford (1975), pioneers in the study of general creativity, pointed out that the creative act needs something to be creative about. In other words, the first step is to perceive, **identify** and clarify a problem, or something which needs explaining or solving. It is seen by some scientists as the most difficult part of the scientific endeavour (Einstein and Infeld, 1938). If there is interest in this space, non-algorithmic problems tend to be provided for the student to solve as this can make subsequent activity more contained and manageable, particularly when it is intended to lead to practical activity. If, however, students are to experience creative thinking in chemistry in a fuller way, the problem space should not be ignored as there can be occasions when students can find their own problems to explore, either partially or in their entirety. Even young students can ask interesting questions (and sometimes fundamental ones), some of which may lead to classroom investigations. These questions, however, are not always the first ones they ask. Often, they begin with factual questions as they clarify and grasp the nature of the phenomenon, and then move on to questions calling for explanations. This means that teachers need to allow students time to ask these and then give careful attention to what follows them (Newton *et al.*, 2018).
    - *Why does bread change colour when it's toasted? What will happen to the copper sulphate crystals if we add another sulphate to the solution before it evaporates? Why do some things dissolve in water when others don't? Why are some things sticky? I read that they used to make something to wash clothes in by soaking wood ash in water: Why did that work? When a large spoonful of common salt is slowly added to a full glass of water, the water does not overflow: Why? If we put exothermic reactions in order from most to least, will there be a pattern to the list?*
- ***The hypothesis space***
  - Having found or been given a problem – perhaps an unexpected observation, phenomenon, or event – the student constructs (*i.e.* **creates**) and proposes a tentative explanation of it. Such an explanation typically rests on a provisional understanding, that is, a mental model of how the components might interact under the given conditions (Klahr and Dunbar, 1988; Newton, 2010). This, in turn, is the product of the student's prior knowledge and imagination which brings ideas together and articulates them. This proposed

explanation has to have some plausibility, at least from the point of view of the students, but that is not to say it must be 'correct' in the sense that it is the prevailing view of chemists in general: an explanation can be novel, plausible, and economically elegant (the three attributes of creativity listed above) and, when tested, still be found to be inadequate. The goal here is not necessarily to replicate the history of chemistry (although that can happen); rather it is to practise and keep alive the student's creative thinking. Surprisingly, it is an aspect of creative thinking in science which may pass unrecognised in the classroom in the eagerness to move on to a more generally recognised opportunity.

  - *I think kitchen foil is a metal because it is shiny like all metals. I don't think air dissolves in water because it isn't mixed in with it. I wonder if an antacid tablet settles your stomach because it neutralises the hydrochloric acid. I wonder if Cola™ cleans copper coins because it dissolves copper oxide. I think that the black residue we get when we heat organic substances is carbon because they all contain carbon and soot is carbon. I think the density of the Dead Sea is increasing because the water is evaporating and not being replaced. I think the properties of an alloy will be a combination of the properties of the elements in it.*

- ***The experimental space***
  - Here, the **creative design** of an appropriate practical test of a hypothesis is the next opportunity for creative thinking (Newton, 2010). Applying some imagination to bring together knowledge and know-how to see if a prediction will survive a reality check again calls for appropriate, novel thinking, and is all the better if it is elegantly simple. This aspect of chemistry teaching is often recognised as an imaginative, creative process, and it is not uncommon for the critical evaluation of such tests to be a significant exercise. Some chemists may specialise in one of the creative spaces and see themselves as Theoretical Chemists or Practical Chemists. Students, however, may benefit from a broad experience, at least in the earlier part of their learning.
    - *Some of the above explanations may be refined and lend themselves to testing by an appropriately designed experiment (and some may be found wanting). For example, chemical properties of an alloy like brass may be compared with those of zinc and copper; an attempt may be made to dissolve copper oxide powder in Cola™; and the effect of antacid tablets on dilute sulphuric acid may be observed.*
- ***The application space***
  - Where there is no wish to divide chemistry into so-called Pure and Applied, the opportunities for creative thinking can be extended into the application of chemistry to solve practical problems (Newton, 2010). Once again, the problem may be found by the student or given. There is usually a need for clarification, followed by an attempt

to construct (**create**) a solution. The first attempt may be found to be ineffective or otherwise inadequate and further exploration, designing, testing and evaluation is needed so the process is, in essence, an iterative one. Of course, existing knowledge alone may be insufficient to solve the problem and, in practice, the creative thinking in the application space may also involve creative thinking in the above three spaces.

    - *Can you use your knowledge of chemistry to create a safe, cheap, effective window cleaning liquid? How might we make 'fog' to add atmosphere to what is supposed to be a haunted house? How might the greenhouse gas, carbon dioxide, be extracted from the atmosphere and made into a useful product at low cost? (*Li and Anastas, 2012*). Using data about readily available substances on Mars, consider the feasibility of 'terraforming' it, that is, giving it a breathable atmosphere.*

- ***Auxiliary spaces***
  - On occasions, activities may be provided which contribute to or develop competence in the above spaces, although may not, in themselves, lead to a full opportunity in any one field. For example:
    - *What if? thinking in which some practicality is suspended just for the fun of exploring the consequences of crazy hypotheses. For instance, what would happen if you built a Periodic Table out of bricks made of each element? (*Munroe, 2014*; Munroe's examples illustrate very well the need to explore the meanings of What if questions.)*

To foster such thinking, teachers need to construct facilitating contexts and strategies which provide opportunities for students to develop competence in these spaces. There is no algorithm for constructing such opportunities (and if there was, robots could do it), but it helps to think in terms of the students' psychological needs, their hopes, and what they see as relevant to their lives (and therefore of interest). Of these, there can be a common need to relate to others so tying what is to be taught to individual people and society in general can facilitate the process of lesson construction. Equally, novelty and surprise can add effectiveness. Nevertheless, students are different, they change as they mature and, because the world changes, what works for one generation may not work for another (Newton, 1988). There are, however, general strategies or devices which lend themselves to diverse uses. For example, students may be set a challenge or problem to solve as a project, and present their findings in a Science Fair, or as a Poster. (*We have no paint to use in art lessons. Can you use the chemicals in vegetables to make some paints? Might it make a difference if you use the vegetables raw or if you cook them first? Why do you think that? Would adding a few drops of cooking oil? What makes you think that? How will you test your ideas?*) Science Fairs, Posters and Problem Solving are relatively common devices for supporting teaching and learning and may be applied widely, but, like everything else, variety in teaching can alleviate a resentful, "Another bloody problem!"

Thinking in the creative spaces is more likely to be productive with an extended engagement. At a time when students' attention span seems to be declining because of their immersion in a deluge of information from the digital world (Kandel, 2006), whatever fosters an engagement is to be valued. Some contexts are illustrated later, and within these there may be strategies which increase the likelihood that students will think in the required way. For example, focused questioning is a technique which supports progress in a student's thinking by guiding it to one of the above spaces, exploring and refining ideas. This can include Socratic questioning emphasising 'Why?' questions, more or less obliging the student to construct an explanation which may, later, lend itself to practical testing (Newton and Newton, 2018). 'Fertile' questions can also provide a starting point for explorative and extended study (Harpaz, 2005).

Not all frames of mind favour thinking creatively, either for the teacher or the student. Time pressure, for instance, tends to switch the mind to conscious, rule-based, thinking, when what is wanted is a frame which favours a more relaxed kind of thought which allows associations to be made. In the same way, high stakes examinations which generate strong anxiety can make creative thinking less likely. It is possible to make creative thinking unlikely merely through the order of events. Students who do not know one another tend to be in a guarded frame of mind when they first meet. Setting them a task which calls for imagination and creative thinking at the outset in the belief that this will ease the tension is likely to lead to disappointing outcomes for everyone. In short, there is a need for psychological safety in which there can be some motivation: autonomy, false starts, playing with ideas, and failure without blame support this kind of purposeful thought. This calls for both intellectual and emotional engagement (McLaren, 2012; Newton, 2015).

Creative and critical thinking are often linked, and creative ideas do, at some point, need to be evaluated or subjected to some kind of quality control. Some students can be frozen into inactivity by a tendency to be critical too soon, either by themselves, or by others observing their work. There needs to be some acceptance that, to begin with, it is often necessary to suspend disbelief until an idea has taken form and its potential explored. It is unfortunate that the word 'critical' can have connotations of obligatory negativism. Bertrand Russell's notion of constructive doubt is better as it offers a more balanced evaluation from a positive frame of mind (Newton, 2016; Newton and Newton, 2018). At times, creative thinking may result in what can be a purposeless mind wandering. However, not all mind wandering is unproductive. Thought allowed to wander from and around the problem in hand can make associations which produce ideas that solve the problem. In this sense, mind wandering can have its uses (Irving, 2016).

Finally, when creative thinking is rewarded, it signals to student and teachers that it is worth their time and effort. If teachers give time to fostering creative thinking, time which could be used to support learning of a kind which brings examination success, it is unlikely that opportunities for practising it will receive little more than an occasional exposure. Creative thinking in the sciences is not easy to assess, but it can be done, as will be described in Chapter 10, and could have a part in examinations. But creative thinking is about thinking more deeply about chemistry and understanding it and its processes. This, in itself, offers support for examination success when these examinations go beyond tests of rote learning.

1.5 Some Challenges

What would foster creative teaching? On the one hand, there are personal beliefs and qualities which can predispose a teacher to imagine 'What if?' scenarios in conjunction with pedagogical theories and an understanding of students. For example, amongst such beliefs are conceptions of teaching; if creative teaching is seen as a valued and expected part of the teaching role, and if there is satisfaction in it, a teacher will be more likely to explore beyond existing practice (*e.g.* Grube and Piliavin, 2000). This is even more likely if teachers have experience of creative activity and confidence in their own creative competence (Lee and Kemple, 2014). On the other hand, such beliefs and attitudes may be undermined by an educational environment that allows little teacher autonomy, where past practices are the only bench mark of success, where teachers are judged only by their ability to reproduce traditional lessons, and where there is no time for imaginative experimentation (Cropley, 2014; Davies *et al.*, 2017).

There can also be some misconceptions about creative thinking which misdirect planning and how others perceive intention. Words like *creative thinking* can have a variety of meanings in everyday life, and have beliefs attached to them which are sometimes carried over unhelpfully into education. Here are some:

- In its everyday use, *creativity* tends to be used to describe a wide range of craft activity, including copying models and following step-by-step instructions or patterns. That is, it includes reproductive making. This can be a rewarding activity but, here, creativity requires some element of novelty.
- There is a tendency to talk of the 'creative arts' and, when exhorting educators to give more attention to creativity, this often means increasing the emphasis on these areas of the curriculum. People commonly place the sciences and mathematics at or near the bottom of the list as far as creativity is concerned. We argue that there can be creative thought and action in almost any kind of human endeavour and, in particular, right across the curriculum (Newton, 2012). At the same time, while there may be some who are broadly creative, whatever the domain, we see it as quite possible that a student may be creative in one domain but not in another. The nature of the opportunities may vary as the disciplines change, and what is meant by novelty, appropriateness, and quality may have different nuances.

- Creativity is often seen as the preserve of the genius, sometimes cloistered in cold garrets, producing objects of such brilliance that their true nature may not be recognised by contemporary society. Creativity and intelligence are not the same concepts; being highly intelligent does not necessarily imply also being highly creative. This is not, however, to say that ignorance is an advantage; knowledge and know-how (or knowing what is needed and how to acquire it) helps (Simonton, 2017). Boden (2009) points out that creative thinking is an aspect of normal, mental activity, something we engage in daily in many small ways that pass unnoticed.
- Following this, some might argue that their students do not know enough to be creative. But what is needed is to know enough to be creative with the given task or problem. To use an everyday example: fat is an ingredient of biscuits, but someone finds there is insufficient fat. Knowing that chocolate has a high fat content, it is used to make up the deficit and save the day. Boden (2001, p. 102) argues that creativity is grounded in knowledge, and 'educators must try... to nurture the knowledge without killing the creativity'. It is easy to give attention to the accumulation of knowledge without giving too much thought to how it might be used.
- Some say that being creative is too difficult for their students. After all, professional chemists spend their lives trying to create something new, and they do not always succeed. A distinction needs to be made between bringing something new to *everyone* into the world, (sometimes called Big C creativity), and bringing something new to the *student* into the world (little c creativity) (*e.g.* Amabile, 1983; Boden, 2004). Dalton's notion of the atom was of the first kind, while a student's notion is more likely to be of the second. Imagination and association of ideas were involved in both, and, while the student may not shake the foundations of chemistry, he or she exercised the creative processes.
- Occasionally, we find that students believe that only virtuous, worthy people are creative. The Italian goldsmith and sculptor, Benvenuto Cellini (1500–1571), created unsurpassed works of art, yet murdered a rival. Today, internet fraud often calls for a high degree of creative thinking, but put to criminal ends.
- Some may also believe that creativity itself is always a good thing. Craft has argued that there is a dark side to creativity when it produces continual change, unsettling and demoralising those subject to it (Craft, 2006). At the same time, not all societies feel comfortable with someone's overt, creative expression. Cropley (2004) argues that creative thinking is usually celebrated in cultures which value individuality but may be concealed in some communal cultures (Cropley, 2004).

Even where what counts as creative teaching and creative thinking in chemistry is occurring, teaching creatively can be thwarted by programmes with a strong emphasis on information acquisition, or which offer little time for experimenting with teaching. Like all creative thinking, it does not come with a guarantee of complete success, and so may need time for further development. Nevertheless, even in adverse circumstances, it can happen, particularly where the teacher has a passion for the subject (Craft *et al.*, 2014). It may even refresh or generate that passion. The potential rewards in student learning and in job satisfaction are considerable (Schacter *et al.*, 2006). That passion, however, can be dampened by an environment that places creative competence low on the list of worthwhile goals, and when tests and examinations do not reward it.

There are also challenges to be met if students' creative thinking is to be exercised in the classroom in chemistry. The myths are very prevalent and believed. Some subjects may be seen by students as nothing more than an accumulation of facts. Perhaps past experience of how these are taught and the nature of examinations have reinforced that belief. Even in universities, the so-called 'cities of ideas' where the rhetoric seems to favour innovation and creativity[†], in practice, graduates are turned out as from a conveyor belt (McLaren, 2012). This is not confined to the UK and has been noted elsewhere in Europe (such as by Semmler and Pietzner (2017) where chemistry teachers say they value creative competence yet are uncertain about its meaning in the context of chemistry). Given this, it could help to highlight some opportunities for creative thought in chemistry, showing what makes them worthwhile with the expectation that chemistry teachers can adapt, extend, and develop opportunities themselves.

Being able to teach creatively and foster creative thinking has probably always been worthwhile – simply repeating the same lessons year after year may never have been what is best for students. Robots could recycle lessons in this way, and even offer variations of them, but students are complex. At least at present, a robot's ability to adapt to the needs of the particular students in front of it has limits. These students are not the 'average' student, nor are they like those of yesterday, or those of tomorrow. The ability to detect the difference, accommodate it, and make the most of it through creative teaching, is probably the most valuable expertise the teacher can bring to the chemistry classroom.

References

Acar S., Burnett C. and Cabra J. F., (2017), Ingredients of creativity, *Creat. Res. J.*, **29**(2), 133–144.

Amabile T. M., (1983), *The Social Psychology of Creativity*, New York: Springer.

Bakshi H., Frey C. B. and Osborne M., (2015), *Creativity vs. Robots. The Creative Economy and The Future of Employment*, London: Nesta.

Ball P., (2006), What chemists want to know, *Nature*, **442**(3), 500–502.

[†]Innovation goes beyond creativity in that it is the successful exploitation of new ideas (Gault, 2018). Creative thinking in chemistry may lead to innovation.

Black A. E. and Deci E. L., (2000), The effects of instructors' autonomy support and students' autonomous motivation on learning organic chemistry, *Sci. Educ.*, **86**(6), 740–756.

Boden M., (2001), Creativity and knowledge, in Craft A., Jeffrey B. and Leibling M., *Creativity in education*, London, Bloomsbury, pp. 95–102.

Boden M. A., (2004), *The Creative Mind: Myths and Mechanisms*, London: Routledge.

Boden M., (2009), Computer models of creativity, *AI Magazine*, **1**, 23–34.

Claxton G., (2006), Expanding the capacity to learn: A new end for education, *Opening Keynote Address, British Educational Research Association Annual Conference*, **vol. 6**, pp. 1–19.

Craft A., (2006), Fostering creativity with wisdom, *Camb. J. Educ.*, **36**, 337–350.

Craft A., Hall E. and Costello R., (2014), Passion: Engine of creative teaching in an English University, *Think. Skills Creat.*, **13**, 91–105.

Cropley A., (2004), Creativity as a social phenomenon, *Creat. Cult. Diversity*, 13–23.

Cropley A., (2014), Neglect of creativity in education, in S. Moran, D. H. Cropley and J. C. Kaufman (ed.), *The Ethics of Creativity*, New York: Palgrave Macmillan, pp. 250–264.

Darby L., (2005), Science students' perceptions of engaging pedagogy, *Res. Sci. Educ.*, **35**, 425–445.

Davies L., Newton L. and Newton D., (2017), Creativity as a 21st Century Competence: An Exploratory Study of Provision and Reality, *Education*, 3–13.

Dijrksterhuis A. and Nordgren L. F., (2006), A theory of unconscious thought, *Perspect. Psychol. Sci.*, **1**(2), 95–109.

Einstein A. and Infeld L., (1938), *The evolution of physics*. New York, Simon & Schuster.

Gault F., (2018), Defining and measuring innovation in all sectors of the economy, *Research Policy*, **47**(3), 617–622.

Gibson R., (2010), The art of creative teaching, *Teach. High. Educ.*, **15**(5), 607–613.

Grube J. A. and Piliavin J. A., (2000), Role identity, organizational experiences, and volunteer performance, *Pers. Soc. Psychol. Bull.*, **26**(9), 1108–1119.

Guilford J. P., (1975), Creativity, in I. A. Taylor and J. W. Getzels (ed.), *Perspectives in Creativity*, Chicago: Aldine, pp. 37–59.

Harpaz Y., (2005), Teaching and learning in a community of thinking, *J. Curric. Superv.*, **20**(2), 136–157.

Hirsch E. D., (2017), *Why Knowledge Matters*, Cambridge, Mass: Harvard Educational Press.

Holbrook J., (2005), Making Chemistry teaching relevant, *Chem. Educ. Int.*, **6**(1), 1–12.

Irving Z. C., (2016), Mind wandering is unguided attention, *Philos. Stud.*, **173**, 547–571.

Jeffrey B. and Craft A., (2004), Teaching creatively and teaching for creativity: distinctions and relationships, *Educ. Stud.*, **30**(1), 77–87.

Kandel E., (2006), *In Search of Memory*, Norton: New York.

Klahr D. and Dunbar K., (1988), Dual space search during scientific reasoning, *Cogn. Sci.*, **12**, 1–48.

Koestler A., (2017), *The Sleepwalkers*, Penguin: London.

Lee I. R. and Kemple K., (2014), Preservice teachers' personality traits and engagement in creative activities as predictors of their support for children's creativity, *Creat. Res. J.*, **26**(1), 82–94.

Lewis S. E., Shaw J. L. and Heitz J. O., (2009), Attitude counts: self-concept and success in General Chemistry, *J. Chem. Educ.*, **86**(6), 744–749.

Li C.-J. and Anastas P. T., (2012), Green Chemistry: present and future, *Chem. Soc. Rev.*, **41**, 1413–1414.

McLaren I., (2012), The contradictions of policy and practice, *London Rev. Educ.*, **10**(2), 159–172.

Munroe R., (2014), *What If?* London: John Murray.

NACCCE, [National Advisory Committee on Creative and Cultural Education], (1999), *All Our Futures: Creativity, Culture and Education*, London: Department for Education and Employment.

Newton D. P., (1988), *Making Science Education Relevant*, London: Kogan Page.

Newton D. P., (2010), Assessing the creativity of scientific explanations in elementary science, *Res. Sci. Technol. Educ.*, **28**(3), 187–201.

Newton D. P., (2012), *Teaching for Understanding*, London: Routledge.

Newton D. P., (2014), *Thinking with Feeling*, London: Routledge.

Newton D. P., (2015), There's more to thinking than the intellect, in R. Wegerif, J. Kaufman and L. Li (ed.), *The Routledge International Handbook of Research on Teaching Thinking*, London: Routledge, pp. 58–68.

Newton D. P., (2016), *In Two Minds: The Interaction of Moods, Emotions and Purposeful Thought in Formal Education*, Ulm-Germany: The International Centre for Innovation in Education.

Newton L. D., (2012), Creativity for a New Curriculum: 5-11, London: Routledge.

Newton L., (2017), *Questioning: A Window On Productive Thinking*, Ulm: International Centre for Innovation in Education.

Newton L. D. and Newton D. P., (2018), *Making Purposeful Thought Productive*, Ulm: ICIE.

Newton D. P., Newton L. D. and Abrams P., (2018), Students' questioning and problem Finding, 16th International Conference on Excellence, Creativity and Innovation in Basic-Higher Education and Psychology, ICIE, Paris, 3 July–6 July 2018.

Pollard V., Hains-Wesson R. and Young K., (2018), Creative teaching in STEM, *Teach. Higher Educ.*, **23**(2), 178–193.

Rinkevich J. L., (2011), Creative teaching: Why it matters and where to begin, *Clearing House*, **84**(5), 219–223.

Rosch E., (1999), Principles of categorization, in Margolis, E. and Laurence, S., *Concepts*, MIT Press, pp. 189–208.

Runco M. A., Illies J. J. and Eisenman R., (2005), Creativity, originality, and appropriateness, *J. Creat. Behav.*, **39**, 137–148.

Schacter J., Thum Y. M. and Zifkin D., (2006), How much does creative teaching enhance elementary school students' achievement? *J. Creat. Behav.*, **40**, 47–72.

Said-Meturaly S., Kyndt E. and Van den Noortgate W., (2017), Approaches to measuring creativity: as systematic review, *Creativity*, **4**(2), 238–275.

Semmler L. and Pietzner, (2017), Creativity in Chemistry class and in general - German student teachers' views, *Chem. Educ. Res. Pract.*, **18**, 310–328.

Shaheen R., (2010), Creativity and education, *Creat. Educ.*, **1**(3), 166–169.

Simonton D. K., (2017), Big-C versu Little-c creativity, in R. A. Beghetto and B. Sriraman (ed.), *Creative Contradictions in Education*, Berne: Springer International, pp. 3–19.

Taannaeifer M. R. and Motaghedifard M. M., (2014), Subjective well-being and its sub-scales among students, *Think. Skills Creativity*, **12**, 37–42.

Torrance E. P., (1999), *Creativity in the Classroom*, Washington: National Education Association.

UNESCO, (2001), *Cultural Heritage, Creativity and Education For All*, Paris: United Nations Educational, Scientific and Cultural Organisation.

CHAPTER 2

Creative Thinking

In Chapter 1 we established that creative thinking involves producing something that is both novel and appropriate. In this chapter, we shall explore how the different aspects of creative thinking can be developed within the chemistry classroom. We describe some kinds of purposeful thought that contribute to creative thinking both by the teacher and the students including support for the core creative activity of thinking in chemistry. This chapter is not arguing for a restructuring of curricula to focus on creativity or scientific process but rather to illustrate how creative thinking can be incorporated in and support existing curricula. It will provide guidance and activities on how to embed creative thinking within teaching. We shall also discuss the characteristics of a creative teacher, how to develop a creative mind set and make it part of teacher identity.

2.1 Chemical Thinking

Chemistry curricula have demanding expectations of their students. They may be required to develop meaningful understanding of core concepts and scientific reasoning, apply them to investigate and explain natural phenomena, propose and justify hypotheses, critically analyse experimental data, build arguments based on evidence and discuss societal issues to make informed decisions about their everyday lives. In chemistry, this requires the development of specific ways of thinking and reasoning. Talanquer and Pollard (2010) coined the term "chemical thinking" to encapsulate the knowledge, reasoning and practices that characterise chemical enterprise (the analysis, synthesis and transformation of matter for practical purposes).

Analysis involves applying strategies for detecting, identifying, separating and quantifying chemical substances (Enke, 2001). Synthesis requires the design of new substances and reactions and transformation relates to controlling chemical processes (NRC, 2003). Creative thinking is important for

Advances in Chemistry Education Series No. 4
Creative Chemists: Strategies for Teaching and Learning
By Simon Rees and Douglas Newton

Published by the Royal Society of Chemistry, www.rsc.org

each of these core practices. Therefore, creative thinking is a valuable skill for chemical scientists and the study of chemistry also provides excellent opportunities to develop creative thinking. From the teachers' perspective, developing an understanding of creative thinking, therefore, leads to teaching activities that embed these ways of thinking. Creative thinking involves divergent, convergent, associative and lateral thinking.

2.2 Divergent Thinking

Divergent thinking is a thought process to generate multiple ideas and connections between cognitively distant ideas in response to a stimulus. The generation of ideas is central to progress in science and it is important to foster a learning environment that promotes idea generation. Students may make all sorts of creative suggestions to explain natural phenomena (see Activity 2.1) which may or may not be scientifically accurate. However, as Taber (2016) argues, when students make these kinds of suggestions it can be all too easy to dismiss them as getting the science wrong and inadvertently stifle creative thought. We should make the most of such opportunities to value ideas and encourage the students' active engagement (this is, of course, often easier said than done). Karl Popper (1992) argued that science proceeded through bold ideas that were a departure from received scientific ideas, only some of which will prove to be useful.

There are several components of divergent thinking as originally described by Guilford (1950). These are: fluency (generating many ideas), originality (producing novel ideas), flexibility (producing varied ideas) and elaboration (producing detailed ideas). Divergent thinking is often promoted in business enterprise settings with activities such as "how many different uses can you think of for a paper coffee cup?" but it is equally feasible to apply similar activities in a chemistry context. These opportunities, however, may be overlooked as divergent thinking is rarely assessed. Indeed, if it was assessed in a traditional examination format then the essence of the creative nature of the activity may well be lost as students would be taught common examples that are used and rote learn responses. However, developing divergent thinking is a valuable skill to be promoted and can lead to improved interest, engagement and understanding regardless of final assessment methods. When facilitating a classroom activity to promote divergent thinking (see Activity 2.1), it is important that there is no restriction placed on the suggestions made in terms of their feasibility or practicality.

Activity 2.1 – Divergent Thinking

This activity is designed to use a natural phenomenon as a stimulus for divergent thinking and also explores scientific understanding.

i) Think of a natural phenomenon relevant to an area of the curriculum. This may be an observation such as why is copper used for wires rather than other metals? Or it could be more of a problem-based phenomenon such as: A coin is frozen in an ice cube – write down as many ways as you can think of to release the coin from the ice (modified from Johnstone and Al-Naeme, 1995).

ii) Fluency – groups of students generate as many ideas as possible to explain the observation or solve the problem.

iii) Originality and flexibility – these ideas can be reflected on in terms of how novel (originality) and varied (flexibility) they are. Have the students generated ideas drawing on a range of different physical and chemical principles?

iv) Elaboration – can the students provide more detailed explanations of their ideas? For example, can they elaborate on why copper is a good conductor of electricity? This provides the opportunity to explore the extent of scientific understanding.

Note:

This activity can be made more demanding by extending the question and challenging ideas (for example, aluminium costs less than copper and is used in power lines so why is it not used in all wires?) or by removing more obvious options (*e.g.* you are not allowed to break the ice or heat it with a Bunsen burner or other heat source).

2.3 Convergent Thinking

Convergent thinking is a tightening of the mind to reflect narrowly on an idea and, in one sense, is the polar opposite of divergent thinking. It describes situations such as where questions have only one correct answer and science curricula have, for a long time, been criticised for emphasising this method of thinking (Hudson, 1967). However, convergent thinking is an important component of creativity particularly in relation to problem solving. In this sense, convergent thinking is akin to critical thinking and involves evaluation, selection and refinement of the best ideas. Critical thinking tends to have only negative connotations. Here, however, it involves looking in a more balanced way at an idea, seeking to identify its good and bad aspects. To avoid negative connotations, evaluative thinking may be preferred. Some time ago, Osborn (1957) proposed the Osborn–Parnes Creative Problem Solving Model where problem solving involves several steps. At the initial stage of the creative process, divergent thinking encourages all ideas to be collected and can lead to novel and entertaining ideas even if completely impractical. Convergent thinking is then used to evaluate the suggestions to find the best possible solutions. Therefore, convergent thinking is associated with critical thinking in evaluating the best options from the full range of ideas generated. For example, looking at the suggestions from Activity 2.1, rank the ideas from best to worst. What criteria have you used to do this? How could different criteria be used to obtain a different outcome? Creative problem solving is characterised by alternating between divergent and convergent thinking modes. Initiating critical and evaluative thinking too soon can inhibit, even stop, the creative process.

2.4 Associative Thinking

Associative thinking is the process of linking one thought idea with another. This can lead to novel and useful ideas with connections being made between hitherto unconnected areas. It is also the dominant thinking strategy for memory techniques (Brown, 2007) and is applied in a chemistry context in Activity 2.2. Some may think that they have never needed to use a strategy such as this to learn chemistry content so what is the point? However, in order, to broaden engagement with chemistry, it is important to use creative strategies that widen participation and facilitate meaning and learning. Visualisation is one of these and is something neuroscience recommends to support learning (Howard-Jones, 2014).

Activity 2.2 – Associative Thinking and Memory Skills

Some aspects of the chemistry curriculum require students to memorise information such as particular chemical equations. Below, for example, are a series of ionic equations that undergraduates are required to recall in the first year of a chemistry degree at a UK university. These are a selection from 16 ionic equations that are required. One strategy to help memorise such information is through associative thinking.

i) For the next 60 seconds try to memorise the equations. Then spend five minutes doing something else before trying to see what you can remember.

$$O_2 + 4H^+ + 4e^- \rightarrow 2H_2O$$
$$BrO_3^- + 6H^+ + 6e^- \rightarrow Br^- + 3H_2O$$
$$ClO^- + H_2O + 2e^- \rightarrow Cl^- + 2OH^-$$
$$MnO_4^- + 2H_2O + 3e^- \rightarrow MnO_2 + 4OH^-$$

ii) Try to recall the equations tomorrow or a week later. Even as a subject expert, conversant with the specialist language of chemical equations, you will probably have difficulty recalling all these equations. Imagine what it must be like for students trying to understand these unfamiliar and abstract equations.

The human brain makes strong use of imagery – how often do we say we can remember a face but not the name? Associative thinking can involve linking an image with each of the symbols to make them more memorable. The more creative and outrageous the images the better! For example, electrons (e^-) may be represented as electric eels, H^+ ions as party hats, O_2 as the shape of a person's mouth breathing or making an "oo" sound, and H_2O as a water fountain.

Therefore, the first equation is visualised as a person with four party hats on going "oo", surrounded by four electric eels and walking through a doorway towards two fountains. Walking through the doorway signifies the arrow in the chemical equation. Try developing your own images for each of the symbols and then memorising the equations and recalling them as before.

A different aspect of associative thinking is finding connections between different parts of the curriculum or with other subjects. Students often struggle to make these connections and see information in isolation. It is, therefore, important that we teach our students how to develop their associative thinking skills (Activity 2.3) and start to see new connections within their knowledge.

Activity 2.3 – Linking across the Curriculum

This activity is designed to develop associative thinking skills and recognise links between different curriculum areas.

i) Think of several different areas of the curriculum and challenge yourself or a group to find connections between them. For example, what are the links between amount of substance and chemical bonding? Just asking these questions encourages thinking about these concepts in different ways. For example, size of atoms is relevant to both concepts, however, in terms of amount of substance, mass is significant whereas with chemical bonding, atomic radius is significant.
ii) Now make connections with new curriculum areas. For example, what connections are there between chemical bonding and acids and bases? This may lead to discussions about the relationship between bond strength and acid strength (comparing hydrofluoric and hydrochloric acid, for example).
iii) Divergent thinking – the generated associations can also be assessed in terms of the quality of divergent thinking (see Activity 2.1). To what extent are the ideas novel and varied?

Challenging students to find these links at the end of a topic is a good way to consolidate learning and deepen subject knowledge.

2.5 Lateral Thinking

Logical thinking is concerned with problem solving through a series of vertical steps that follow on from each other in a correct sequence

(De Bono, 2016). This is the predominant approach applied to problem solving in chemistry whether it be mathematical problems, reaction mechanisms or instrumental analysis. Lateral thinking, however, is taking a "horizontal" approach to problem solving to identify new possibilities, often referred to as "thinking outside the box". In lateral thinking, seemingly wrong turns may achieve the solution. In logical thinking this would not apply. Irrelevant information may ultimately be useful whereas logical thinking requires only what is relevant. Therefore, for students who may struggle with applying logical thinking skills, providing opportunities to develop lateral thinking skills can be engaging. Vertical thinking develops the ideas generated by lateral thinking. De Bono (2016) describes it as "you cannot dig a hole in a different place by digging it deeper. Vertical thinking is used to dig the same hole deeper. Lateral thinking is used to dig a hole in a different place". Lateral thinking is a way of using the mind, and techniques are used to make it a habit of mind just as logical thinking is also developed. The mind handles information by forming representations that become established ever more rigidly over time (Newton, 2000). Information that is used as part of one pattern cannot easily be used as part of a completely different pattern. The purpose of lateral thinking is to restructure patterns, putting information together in new ways to create new ideas. With vertical thinking an experiment is designed to show an effect whereas with lateral thinking an experiment is designed to develop ideas. Lateral thinking is a provocative attitude to challenge established patterns. In the classroom, therefore, it is important to try to encourage such contributions even if they may, in the first instance, appear to be inappropriate and a potential distraction.

Students often find it difficult to apply their knowledge and transfer learning to unfamiliar contexts. Consider, for example, the following equation and scenario:

$$6CO_2\ (g) + 6H_2O \rightarrow C_6H_{12}O_6\ (s) + 6O_2\ (g) \quad \Delta H = +2879\ \mathrm{kJ\ mol^{-1}}$$

Each year 3.4×10^{18} kJ of solar energy is taken in by all plants on the Earth to make photosynthesis take place. Calculate the mass of carbon dioxide that is removed each year from the atmosphere by photosynthesis on Earth.

Students may struggle to recognise the chemistry knowledge (*e.g.* amount of substance) to apply to this apparently biological context. Developing lateral thinking can help students think more broadly and how to apply their knowledge in different contexts.

De Bono (2016) describes using ambiguous pictures as a mechanism to promote alternative interpretations. Students are presented with a picture that is open to different interpretations and write down what it shows. Variability between individual interpretations shows ways of looking at the picture. Suggestions should not be judged in terms of which are the best and which are unreasonable. The image, for example, may be of a water bubble bursting (Oefner, 2019) with iridescent colours on the surface and a fine spray of water droplets bursting forth. Alternative interpretations could be a jellyfish, a galaxy or ice crystals.

Case Study 2.1 – Creative Chemists

This is the first of several case studies throughout the book featuring creative thinkers in chemistry.

Lee Cronin

"I fundamentally think that the only way to explore reality is through creative thinking"

(Image courtesy of Lee Cronin)

Lee Cronin is Regius Professor Chemistry at Glasgow University. He runs a large research group called complex and digital chemistry (Cronin, 2019). He considers himself the Principal Investigator and the Creative Director.

What is your background that led you to this point?

At school I was always disruptive and was in the learning difficulties group. I saw the world in my own way and my teachers did not think I was academic even though I liked science and maths. In fact, I'd been doing my own experiments and exploring reality since I can remember from six years old. I made it to A levels and was able to go and do chemistry at university. I have always been interested in how life got started, how chemicals can make brains, how time works and so on.

How has creative thinking been important in your career?

Creative thinking has been responsible for all my ideas, this led to a range of experiments I would have never done following conventional thinking and all the discoveries we have made and are making. For example, I wanted to make a von Neumann machine for chemistry and the challenge allowed me to develop the concept of the Chemputer – a universal programmable chemical robot that can make all the molecules in the literature. This required me to invent a theoretical abstraction that could connect the idea of making a molecule to the synthetic approach, and then solidify that by compiling this down to a real machine level. In turn, this was inspired by my desire to make a massively parallel chemical reaction system that can 'search' for the molecules that might have emerged at the origin of life on Earth.

2.6 Strategies to Promote Creative Thinking

There are several established strategies that are used in a wide variety of contexts to promote creative thinking, such as SCAMPER and De Bono's Thinking Hats.

2.6.1 SCAMPER

Often used in business environments and product design, SCAMPER (originally Eberle, 1971) can be successfully modified for the chemistry classroom. SCAMPER is an acronym that describes a structured cooperative learning technique that guides the students through the activity in steps using the corresponding letters:

S refers to **Substitute** – an alternative but equivalent topic is suggested.
C refers to **Combine** – add information to the original topic.
A refers to **Adjust** – identifies alternative ways to present the topic in more flexible ways.
M refers to **Modify** – creatively changes the topic
P refers to **Put to other uses** – identifies applications for the topic.
E refers to **Eliminate** – removes unimportant parts of the topic.
R refers to **Reverse or Rearrange** – evolves a new topic from the original topic.

Activity 2.4 shows how this strategy can be used in the chemistry context. The value of using this strategy in the chemistry classroom is that students can see how they are learning skills and strategies that can be applied in different contexts. It can also be used to generate impromptu questions and to consolidate knowledge. It is not necessary to use all the letters in the acronym, as some may not be relevant to a particular activity but it does raise awareness amongst the students of different ways of developing creative ideas.

Activity 2.4 – SCAMPER

This activity demonstrates how SCAMPER can be used by students when planning or evaluating an investigation.

i) Pose a problem for the students to solve using the SCAMPER technique. For example, how can the rate of reaction between marble chips and acid be increased?

ii) Use the SCAMPER acronym as a framework to generate ideas. For example:

Substitute
What could we swap to make the reaction faster *e.g.* a different acid?

Combine
What factors could we combine to make the reaction faster *e.g.* a higher temperature and stirring?

Adapt
What would happen if we adapted the experiment *e.g.* a larger volume?

Modify
How could the apparatus be modified?

Put to other uses
What other uses could this experiment be used for, *e.g.* as a method to generate carbon dioxide?

Eliminate
What would happen if we removed part of the apparatus, *e.g.* the delivery tube?

Reverse or rearrange
What if the experiment was rearranged, *e.g.* adding the acid to the marble chips rather than the chips to the acid?

2.6.2 De Bono's Thinking Hats

De Bono's six thinking hats (De Bono, 2000) represent different ways of thinking and is a strategy promoted in organisations to develop creative problem solving. It helps students to explore the different types of thinking involved. Each way of thinking is assigned a different colour as follows:

Emotional (Red) – acting on emotions and impressions rather than logical reasons.
Positive (Yellow) – focusing on benefits and constructive ideas.
Critical (Black) – pointing out errors and negative aspects.
Objective (White) – concerned with facts and figures.
Creative (Green) – innovating and taking new approaches.
Big picture (Blue) – thinking about overall impacts.

Highlighting these helps students appreciate the need for diverse but orchestrated thinking. This can be particularly helpful, and challenging, if the students are required to think in ways that are not compatible with their thinking habits. Intuitive (unconscious) thought processes are inevitable – analytical and critical thought can be used to evaluate their suggestions. This technique may be particularly useful for framing discussions around curriculum areas exploring the impact of chemistry in the world such as plastic recycling or air pollution. De Beer and Whitlock (2009), for example, used the technique for framing discussions about Western and traditional medicines. A classroom activity may use all these different types of thinking or it may focus on one or two of them.

2.7 Some Characteristics of the Creative Teacher

In this section we discuss the characteristics demonstrated by a creative teacher. A creative teacher is open to the potential of new ideas and finding new ideas. They seek opportunities to network and engage with other teachers and professionals; fostering these connections can help find new approaches and solutions to problems. A commitment to continuing professional development from both teacher and employer is essential to support this so that individuals can engage with professional networks and colleagues locally and internationally. To engage effectively in these opportunities, a creative teacher must be open to working collaboratively with others to share and develop ideas.

Equally, it is important to find activities that provide time and space to reflect on, clarify and develop ideas such as: walking the dog, running or meditation. Finding space for this part of the creative process can be very challenging in a time-pressured and demanding environment. However, it can help new ideas and perspectives to develop. Having a notebook to hand to write down ideas as they occur is very useful. Ideas may often come and then go before we have had an opportunity to capture them. It does not matter how absurd the idea may turn out to be later. At such divergent moments, it is important to just note the idea down.

Ultimately, however, creativity cannot flourish in a regimented and a risk-averse environment. It is important that creative ideas (those which are both novel and appropriate) are embraced and valued within a supportive culture that values and fosters creativity and innovation.

2.7.1 Creative Teaching

Creative teaching in chemistry may become restricted to particular enrichment activities that are outside 'normal' chemistry teaching and learning. However, creative thinking strategies can be successfully used to develop innovative ideas in the classroom in a more integrated way. How often are innovative strategies tried but then abandoned because the response was chaotic and learning erratic? However, opportunities for divergent and convergent thinking can provide structure to thinking and developing an understanding of how to move forward and develop the strategy successfully. Activity 2.5 is designed to demonstrate how this can be achieved in relation to students undertaking independent investigations. It does not provide an exemplar of how to undertake this activity successfully with students but, rather, it demonstrates how creative thinking can be used by teachers to generate their own ideas.

Activity 2.5 – Independent Investigations

A creative teacher is keen to provide students with the opportunity to design and undertake an independent investigation, but when it is tried the results can be unsatisfactory and the atmosphere chaotic unless there is forethought. This activity is designed to demonstrate how creative thinking can be applied to this situation.

i) **Fact finding**

Divergent thinking – the teacher identifies all possible factors that could be important in relation to students undertaking independent investigations, such as: suitable investigations, prior learning, resources to develop autonomy, classroom management and so on.

Convergent thinking – the teacher seeks information around these topics (elaboration) and then focuses on the most important sources of information for ideas.

ii) **Ideas generation**

Divergent thinking – from this informed perspective, the teacher generates as many ideas as possible to solve the problem.

Convergent thinking – these ideas are focused on the most important and feasible options. Critical thinking skills are used to evaluate the ideas. Ideas are compared with practical constraints that may help to identify a suitable solution.

It is at this stage that SCAMPER may be useful to encourage new approaches. For example, an option may be discounted because it is considered impractical but, perhaps, an alternative way could be found that would help achieve a similar aim.

iii) **Implementation**

Critical/Evaluative thinking – the teacher uses critical thinking skills to evaluate the effectiveness of the strategy and then make adjustments related to implementation to suit their needs.

Alternatively, the curriculum topic can be identified and then creative thinking applied to address it. Activity 2.6 provides an example of how creative thinking is applied in developing pedagogy for individual curriculum statements. Creative teachers develop these creative thinking strategies so that they have a suite of activities that are suitable for a range of contexts and learners. They are responsive and flexible to the learning situation.

Activity 2.6 – Curriculum Goal

This activity demonstrates how creative thinking is applied to develop pedagogical strategies. Consider the following statement made in a UK GCSE chemistry syllabus in relation to intermolecular forces and size of molecules:

"*Substances that consist of small molecules are usually gases or liquids that have relatively low melting points and boiling points. These substances have only weak forces between the molecules (intermolecular forces). It is these intermolecular forces that are overcome, not the covalent bonds, when the substance melts or boils. The intermolecular forces increase with the size of the molecules, so larger molecules have higher melting and boiling points. Students should be able to use the idea that intermolecular forces are weak compared with covalent bonds to explain the bulk properties of molecular substances.*" (AQA, 2019).

i) **What types of thinking are required to meet this objective?**
   In developing a pedagogy to address this statement, a range of creative thinking skills are required such as:
   - Operating at the sub-microscopic level, visualisation and abstract thought are required to visualise the molecular processes.
   - Associative thinking by finding analogies in a more familiar context (macroscopic) could help students relate to the sub-microscopic level. Analogical thinking is powerful and can produce understandings which lead to predictions for practical investigation (Newton, 2003).
   - Divergent thinking promotes ideas generation of analogies.
   - Convergent thinking can be used to review the limitations of the analogies.

ii) **What types of strategies may be developed?**
   These creative thinking processes can lead to the development of a range of strategies that may be useful in different contexts and situations. A creative teacher develops a range of strategies that enables them to be responsive to the challenges they encounter. Example strategies include:
   - **Role play**: students link arms in a line with different numbers of students "bonded" together to represent different sizes of molecules. These "molecules" then approach each other and a piece of string is held between the "molecules" to represent the weaker intermolecular forces. Energy can then be applied by encouraging the students to move more and break the intermolecular forces.
   - **Analogies**: students can be challenged to identify an everyday parallel example that illustrates the same principle such as: interlinking fingers, pages of two books or tangled necklaces.

The strengths and limitations of the different analogies can then be discussed. De Bono's Thinking Hats can be used here to encourage the students to apply one type of thinking at a time.

2.8 Conclusion

In this chapter we have discussed the different aspects of creative thinking, strategies for developing creative thinking and how to apply them within the chemistry teaching context. We have demonstrated a range of different activities that can be used to support learning and develop pedagogical activities. Conscious application of these strategies will help embed creative strategies more routinely within the classroom and is demonstrated in a range of different contexts throughout this book. In the next chapter, for example, we go on to consider creative engagement with all our senses to enhance learning in chemistry.

References

AQA, (2019), GCSE Chemistry, Available at https://www.aqa.org.uk/subjects/science/gcse/chemistry-8462.

Brown D., (2007), *Tricks of the Mind*, Channel 4 Publishing.

Cronin L., (2019), *The Cronin Group*, Available at http://www.chem.gla.ac.uk/cronin/.

De Beer J. and Whitlock E., (2009), Indigenous knowledge in the Life Sciences classroom: put on your De Bono hats!, *Am. Biol. Teach.*, **71**(4), 209–217.

De Bono E., (2000), *Six Thinking Hats*, London. Penguin.

De Bono E., (2016), *Lateral Thinking: A Textbook of Creativity*, London: Penguin Life.

Eberle R. F., (1971), *Scamper: Games for Imagination Development*, Buffalo.

Enke C. G., (2001), *The Art and Science of Chemical Analysis*, John Wiley & Sons Inc, **vol. 1**.

Guilford J. P., (1950), Creativity, *Am. Psychol.*, **5**, 444–454.

Howard-Jones P. A., (2014), Neuroscience and education: myths and messages, *Nat. Rev. Neurosci.*, **15**(12), 817–824.

Hudson L., (1967), *Contrary Imaginations: Psychological study of the English schoolboy*, Harmondsworth, England: Penguin.

Johnstone A. H. and Al-Naeme F. F., (1995), Filling a curriculum gap in chemistry, *Int. J. Sci. Educ.*, **17**(2), 219–232.

National Research Council (NRC), (2003), *Beyond the Molecular Frontier: Challenges for Chemistry and Chemical Engineering*, Washington, DC: National Academy Press.

Newton D. P., (2000), *Teaching for Understanding*, London: Routledge.

Newton D. P., (2003), The occurrence of analogies in elementary school science books, *Instr. Sci.*, **31**, 353–375.

Oefner F., (2019), *Incredible Soap Bubble Bursting Photography*, Available at https://www.designswan.com/archives/incredible-soap-bubbles-bursting-photography-by-fabian-oefner.html.

Osborn A. F., (1957), *Applied Imagination: Principles and Procedures of Creative Thinking* (rev. ed.), New York: Scribner's.

Popper K. R., (1992), *The Logic of Scientific Discovery (reprint)*, New York: Rutledge.

Taber K. S., (2016), 'Chemical Reactions Are Like Hell because…': Asking Gifted Science Learners to be Creative in a Curriculum Context that Encourages Convergent Thinking, in *Interplay of Creativity and Giftedness in Science*, Brill Sense, pp. 321–349.

Talanquer V. and Pollard J., (2010), Let's teach how we think instead of what we know, *Chem. Educ. Res. Pract.*, **11**(2), 74–83.

CHAPTER 3
Multisensory Learning

In the previous chapter, we explored the nature of creative thinking in chemistry and strategies to develop a creative mind set. In this chapter, we focus on multisensory learning as a mechanism for teaching *with* creativity (Beghetto, 2017). As Torrance (1988) said, "*Creativity is almost infinite. It involves every sense – sight, smell, hearing, feeling, taste and even perhaps the extrasensory.*" Teaching creatively means applying principles and techniques of creativity to subject matter teaching. It can be one of the hardest things to achieve as it requires confidence in the strategies to be used, preparedness to take risks and acceptance of failure. This chapter uses specific examples and case studies to help the reader develop understanding of what teaching creatively can look like in this domain.

Humans are multisensory in order to interpret and understand their environment. The brain is attuned to operate optimally by integrating information from all the senses to develop a response. Formal education, however, tends to favour a limited number of senses with a predominance for visual and auditory stimulation. In this chapter, we discuss how pedagogical approaches that encompass a richer sensory experience can help students engage with and understand some of the most challenging and demanding concepts in chemistry education.

Multisensory approaches may traditionally be associated with students with specific educational needs. However, if chemistry educators are to engage a more diverse cohort of students in the subject then these strategies should be embedded far more frequently in chemistry teaching. Multisensory learning is about creating richer learning experiences that enable students to experience chemistry in new ways that help engage, develop understanding of the sub-microscopic level and promote creative thinking.

Advances in Chemistry Education Series No. 4
Creative Chemists: Strategies for Teaching and Learning
By Simon Rees and Douglas Newton

Published by the Royal Society of Chemistry, www.rsc.org

3.1 Teaching with Creativity

Scholars of teaching have identified that teaching any subject matter requires more than just good subject knowledge. Shulman (1987), for example, explained three types of knowledge required for successful teaching: subject matter or content knowledge, general knowledge of teaching strategies (pedagogical knowledge) and more specific knowledge of how to teach particular subject matter to a particular set of students in a particular context (pedagogical content knowledge). Teaching with creativity applies the principles and techniques of creativity to subject content teaching. Beghetto (2017) describes the knowledge required to do this as Creative Pedagogical Domain Knowledge; a blend of creativity-domain knowledge (knowledge of key creativity concepts, theories and studies from the field) and creative pedagogical knowledge (knowing how to teach a particular population of students about creativity in a particular context). This chapter aims to demonstrate how these principles can be applied to teaching chemistry using multisensory experiences.

3.2 Engaging All the Senses for Learning

Typical educational experience is dominated by the visual and oral senses. Students receive written and oral instruction, observe, undertake tasks and provide written and oral responses. However, everyday experience of life is a far more varied sensory experience, be it the smell of fresh coffee in the morning (olfactory), the feel of water on our hands (haptic) or the taste of our favourite meal (gustatory). It is through these varied inputs that we make sense of the world around us. Teaching creatively should make use of all of these senses to engage students in learning about and understanding chemistry. This is particularly important when exposure to the subject is only for an hour or two a week and it is in competition with the rich sensory interactions they experience during the rest of the day. Actors use techniques of emotional memory and recall the smells, touch, tastes sights and sounds associated with a role or situation to engage more deeply with their role (Cox *et al.*, 2016). We argue that there are benefits for learning and engagement to using more of the senses and engaging more deeply with chemistry.

3.2.1 Sensing the Sub-microscopic Level

An important component of pedagogical content knowledge in chemistry is an awareness of the challenges of multilevel thinking. Johnstone's triplet (Johnstone, 1991) explains how chemistry students have to operate between the macroscopic (what you can see and touch), sub-microscopic (atoms and molecules) and symbolic levels (formulae, equations *etc*). Multisensory experiences are readily associated with the macroscopic level. In any experiment, students may describe what they see (a colour change), hear (a pop), smell (a gas produced) or feel (the test tube is warm). However, the sub-microscopic and symbolic levels are far more difficult for students to experience and understand. This section demonstrates how multisensory learning provides students with macroscopic experiences of the sub-microscopic level.

Key Questions

These key questions promote creative thinking about the challenges of teaching about the sub-microscopic level.

1. What are the key challenges for students to understand the nature of matter at the sub-microscopic level?
2. What aspects of creative thinking are the most important to operate at the sub-microscopic level?
3. How can novel multisensory experiences be developed to improve understanding?

3.2.2 Electrostatic Interaction

In a student's everyday experience, objects interact through direct contact forces and the force of gravity. It is only natural, therefore, that this interpretation is transposed to interpret interactions at the atomic level. However, electrostatic forces are dominant at the sub-microscopic level. The challenge for the creative chemistry teacher, therefore, is to develop strategies that enable the students to develop their conceptual understanding of the interactions at the atomic level. In this section we describe how to think creatively about the sub-microscopic level to enhance student conceptual understanding.

3.2.3 Particulate Nature of Matter

Students may struggle with developing an understanding of the particulate nature of matter at the sub-microscopic level (Taber, 2014) as compared to a material's macroscopic properties. Particles, be they atoms, molecules, electrons, protons or neutrons, are typically represented as drawings of circles, dots or crosses (Figure 3.1).

When Lewis (1916) developed representations such as these to describe covalent bonding, they served as tool for the experienced chemist to represent and understand the sub-microscopic world that they have spent a career studying. Mahaffy (2006) describes these representations as our "chemical canvas" which convey great meaning about important molecules in our lives. The novice chemist, however, is approaching these representations in the other direction. They do not have the benefit of years of study to provide insight into their meaning and limitations. These representations may be more readily associated with childhood games such as noughts and crosses

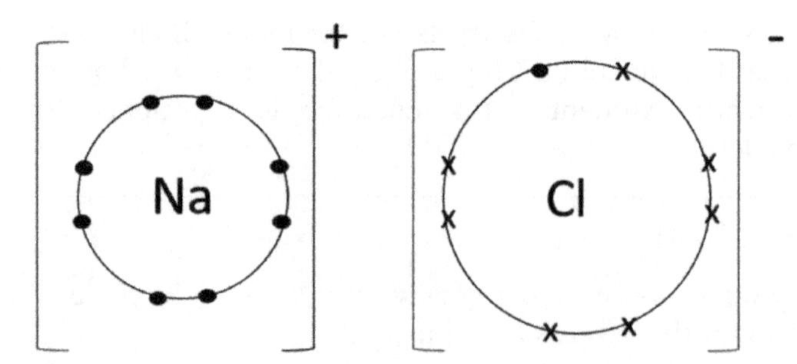

Figure 3.1 Typical representation of ionic bonding in sodium chloride.

(tic tac toe) and there is little sense of the dynamic sub-microscopic world they are representing. Therefore, it is fundamentally important that the chemistry teacher lifts these representations off the page to develop student understanding of this unfamiliar world.

These drawings may be physically represented as hard spheres such as marbles that interact and collide with each other as determined by Newton's laws. This has the sensory benefit of providing something that can be physically observed and handled. However, this representation is very different to the reality of interactions at the atomic level where electrostatic forces are dominant. These representations may reinforce misconceptions about sub-microscopic matter with individual atoms behaving in the same way as macroscopic amounts.

Using creative thinking, how can we develop different sensory experiences to simulate interactions at the sub-atomic level? Activity 3.1 describes an alternative that enables students to observe electrostatic forces dominant over gravitational forces at a macroscopic level.

Activity 3.1 – Electrostatic Dominance

Small polystyrene balls (3–4 mm diameter) used to fill beanbags have an observable size but low mass. When handled, these "particles" demonstrate interactions where electrostatic forces are dominant to the students through visual and haptic stimulation. The students observe the balls behaving differently to what may be expected.

i) Pick up a handful of the balls and then try to tip them back into the container and observe.

   Some of the balls will pour back into the container but others will spill out in an unpredictable way or remain stuck to your hand. Compare this with picking up a handful of dry peas for example.

ii) Place some of the balls in a clear plastic container. Gently agitate the box and observe the particle interactions. Tip the box upside down and observe how some of the balls "defy" gravity and remain stuck to the sides. Compare this with heavier balls such as dried peas or marbles. What differences are observed?

When the balls move and collide they can be observed following paths that are influenced by electrostatic attractions to particles nearby rather than mass.

iii) Charge up a Perspex rod and demonstrate how the electrostatic force can be used to move the particles around and is stronger than the gravitational force.

iv) Add more balls to the container and agitate.

The particles will arrange themselves in a regular pattern characteristic of a solid. Reduce the number of particles to simulate particulate arrangements in liquids and gases.

This representation provides students with a macroscopic physical sensory experience of the dominant force at the submicroscopic level.

3.2.4 Atomic Structure

Multisensory representations are also useful for developing understanding of atomic structure.

Consider the following questions:

- What are the challenges associated with student conceptualisation of atomic structure?
- What strategies do you use to help develop understanding?
- What other strategies may be useful to promote a better realisation of matter at the atomic level?

Typical representations of atomic structure for a student are limited to diagrams consisting of circles and crosses to represent the different parts of the atom. With the crosses representing electrons arranged on concentric circles radiating out from a central circle or circles. How can this representation be improved and provide students with a more tangible sensory experience of what this unobservable world is really like?

The use of physical objects such as counters or playdough that the students can physically handle, assemble and manipulate can provide richer sensory experiences. However, once again, there is no sense of the electrostatic forces prevalent in an atom. Rarely will a student question the assumptions inherent within the model such as why do the protons not repel each other if they are the same charge? Or why do the electrons not move towards the protons if they are oppositely charged? These are not novel questions in the field of human knowledge as they have been asked many times before and answers sought. However, for the individual student learning about these concepts for the first time, it is creative thinking that leads to novel ideas in that context.

The representation leads to limited creative thinking of the model presented. By providing a more multisensory experience using magnetic forces (Activity 3.2), it is possible to create an environment that is more likely to generate creative questions.

Activity 3.2 – Magnetic Modelling of the Atom

Students should be familiar with the properties of magnets and these can be used to consider the atomic model and promote evaluative thinking.

i) Place several magnets on the desk with the same poles facing each other. Try to bring them together so that they are touching. What do you observe? What part of the atom does this represent? What question does this raise about the model of the atom?
ii) Hold one magnet in one hand and then bring the opposite pole of another magnet towards this magnet. What do you feel? Can you rotate the second magnet around the first one so that it does not touch it? What part of the atom is this representing? What question does this raise about the model of the atom?

Through hands on modelling of the atom, limitations of the model may be exposed which promote critical thinking and creative thinking to suggest solutions. Some may argue that students do not need to know why the protons do not repel each other as it is not on the syllabus. However, these strategies are about developing evaluative thinking of the presented model.

3.2.5 Bond Polarity

Beyond the structure of individual atoms, students develop an understanding of the interaction between two atoms as they form a chemical bond. This extends to explaining polarity in covalent bonds due to differences in electronegativity. Students are required to conceptualise these ideas with the aid of diagrams such as in Figure 3.2. Such representations, however, can present significant challenges for novice chemists to interpret at the sub-microscopic and symbolic level. A variety of unfamiliar symbolic language is used to try to convey meaning. Multisensory learning can provide opportunities to model and simulate these phenomena (Activity 3.3).

Activity 3.3 – Bond Polarity and Electronegativity

Electronegativity can be a difficult concept for students and cannot be demonstrated at the sub-microscopic level. However, it can be modelled at the macroscopic level.

i) Arrange two Van der Graff generators approximately 30 cm apart. The spheres represent the two atomic nuclei.

ii) Using thread, suspend two light polystyrene balls or similar (a pair of electrons) midway between the two spheres.
iii) Start charging one of the spheres and the balls can be observed moving towards the sphere or the more electronegative atom. With careful manipulation the "electrons" can be moved towards either of the two nuclei.

In this modelling activity, students experience the effect of electrostatic forces that they can then relate to during their learning. The limitations of this model as a representation of the real situation can also be discussed (*e.g.* the representation of the electrons as discrete particles).

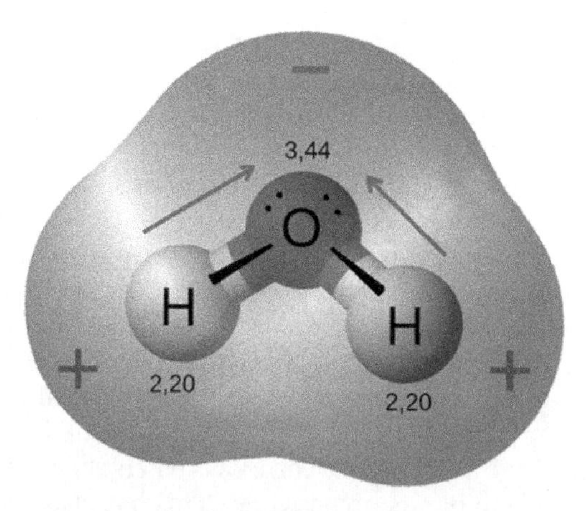

Figure 3.2 A representation of polar covalent bonds in a water molecule. Reproduced from https://en.wikipedia.org/wiki/Chemical_polarity#/media/File:Dipoli_acqua.png under the terms of the Creative Commons Attribution-Share Alike 3.0 Unported licence, https://creativecommons.org/licenses/by-sa/3.0/deed.en.

Case Study 3.1 – Creative Chemists

Cary Supalo

Dr Cary Supalo is the founder of the assistive technology company Independence Science.

Blind since age seven, he obtained his PhD in chemistry from the Pennsylvania State University. He has always had an interest in why things were the way they were and it was this interest in scientific phenomena that initially inspired him to pursue a career in science. However, he encountered active discouragement along the way. One instance that stands out was at a key moment of course selection in secondary school – Cary wanted to take the senior level calculus course so that he could further pursue a STEM degree, whether engineering, computer science or chemistry. He unfortunately was told by teachers and school administration that no blind student had ever taken calculus before. If he chose this path, his school could not support that path. As a result, he had to reluctantly choose a different and much less rigorous maths class.

That summer, in the interlude between graduating secondary school and starting university, Cary attended his first National Federation of the Blind convention. There he was exposed to literally thousands of blind people working in all types of fields including, ironically, a high school calculus teacher. It was here he quickly realized he could pursue any dreams he wanted to – it was just a matter of knowing how to access this information. This grew into a lifelong commitment to fundamental problem solving.

It is this extensive life experience of problem solving that has been an asset to the chemical profession. Cary spent many hours working with undergraduate students who he trained to assist him with his studies. This ranged from reading textbooks and other course handouts, to using proper techniques to communicate maths and chemistry equations, to describing visual aspects of chemistry and other science laboratory activities he was required to perform.

This extensive experience of recruiting and training students to provide him access to information has been a strength throughout his academic and professional career. After graduate school, Cary founded Independence Science, an access technology firm that leverages the experience of Cary and other blind scientists to develop access technologies that provide access to hands-on science learning experiences. Education literature shows how students, in general, learn best when they are given the opportunity to engage in hands-on activities. This has also been found to be true for the blind according to Cary's dissertation work.

After several years working in the private sector and as an entrepreneur, Cary obtained a faculty position as an assistant professor at an American university, where he taught chemistry laboratory courses to sighted students. This experience was very rewarding to Cary and spurred him to pursue a research career that would shift the paradigm for accessibility in education systems. He now works full-time at the Educational Testing Service helping to make high stakes assessments more accessible to students with disabilities. His extensive problem-solving skills now pays dividends to this and other domains he finds himself engaged in.

3.3 Olfactory Learning

In this section, we discuss the potential to engage the sense of smell (olfactory) within creative chemistry teaching. We can all recall occasions in our lives when a particular smell, such as freshly cut grass or shelves of books, elicits strong memories. There is evidence that if someone is able to recollect a smell, then the memories made at that time can be recalled more strongly (Perry *et al.*, 2018). Our sense of smell is linked with an evolutionary drive for food and survival and the identification of harmful or helpful substances.

When we smell something the olfactory bulb in the brain sends information about the smell to other parts of the brain such as the limbic system (responsible for motivation and learning), temporal lobe (explicit memory formation) and the orbitofrontal cortex (reward association) and builds associations between the object and the smell. Thereby, upon detecting the same smell in the future, these associations are recalled. The potential is clear, therefore, to use opportunities within chemistry teaching to use smell to enhance learning.

Key Questions

In what ways do you currently engage students in chemistry through the olfactory and gustatory senses?

Thinking creatively (you may like to use one of the strategies in Chapter 2), how could you make more use of these senses to enhance your teaching?

3.3.1 The Chemist's Spice Rack

Just as the spice rack is essential for a chef to practise his culinary art so should it be an essential addition to every chemistry teacher's repertoire. Student associations with the word "chemicals" is often to think of substances in bottles in the laboratory detached from their everyday experience and should be treated with caution. These perceptions can be challenged when everyday sources of these chemicals are presented in contexts more familiar to the students such as herbs, spices, essential oils and incense. Take a look through your own spice rack and reflect on whether you make use of any of these in your teaching. Do you know about the active ingredients or how you could link these to your teaching? Compound Interest (2019) has produced many useful resources explaining about the structure and applications of these different chemicals. Activity 3.4 explains a simple way to promote students to think creatively about substances and their smells and tastes.

Activity 3.4 – Creative Thinking about the Chemistry of Smell

This activity shows one example of how to incorporate smell in a teaching activity that addresses both chemistry knowledge and creative thinking.

i) Present the students with a range of different mystery smells (*e.g.* turmeric, cinnamon, rosemary, cumin, cloves, vanilla, spearmint, caraway) and ask them to describe the smell and see how many they can identify.

ii) Present the chemical structures of the compounds and ask them to identify any common structures (*e.g.* aromatic benzene rings) and any differences. This is a good opportunity to reinforce organic chemical structures and functional groups.

iii) Promote creative thinking by asking the students to come up with any ideas as to why the different compounds smell different and how the human body actually detects smell.

Notes:

Spearmint and caraway are carvone optical isomers of each other (see learnchemistry (2019)). The detection of smell is a relatively poorly understood phenomenon that provides a useful example of the limitations of human knowledge and developing questions to investigate further.

Rosemary is particularly associated with memory retention. Ancient Greek students wore garlands of rosemary in exams – and Ophelia, in Shakespeare's play Hamlet, says: "There's rosemary, that's for remembrance." Try using this property with the students to design an investigation as to whether it does improve memory. This is an opportunity to reinforce core knowledge as well develop investigative skills.

Provide students with the opportunity to taste some of the herbs and spices and note the connection between taste and smell.

Taking this further, these individual herbs and spices can be the focus of an entire lesson, or for different parts of the curriculum. This section provides two possible examples:

3.3.2 Example 1 – Turmeric

Turmeric is obtained from the root of a plant belonging to the same family as ginger. The root is dried and ground to produce the vibrant orange powder. The spice has a bitter taste and a mustard aroma which is used in a wide variety of Asian cuisine such as curries and also as a dye.

i) Introduction (Structure – functional groups)

Samples of turmeric are passed round for students to smell. It can also be offered to taste with disposable spoons. Students describe the appearance, smell and taste. The structure of curcumin (a key chemical constituent) is displayed and the students are asked to identify key aspects. The structure provides an opportunity for students to identify key functional groups in organic chemistry and reinforce understanding (Figure 3.3). The discussion can be extended to possible tests that could be carried out to identify key functional groups.

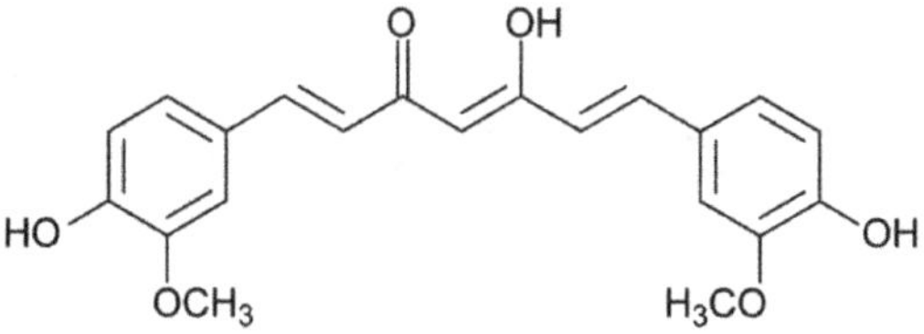

Figure 3.3 The chemical structure of curcumin (alkaline conditions).

ii) Fluorescence – atomic structure

By dissolving turmeric in alcohol (it is insoluble in water) and illuminating with UV light, a bright yellow-green fluorescence is observed. Electrons within the curcumin structure absorb energy from the UV light and move to an excited state. As energy is then lost and the electrons return to ground state, visible light is released. This provides the opportunity to discuss atomic structure and reinforce ideas relating to electrons residing in energy levels and how energy can be applied to move them between levels.

iii) pH indicator

Curcumin's chemical structure changes in acidic or alkaline conditions enabling it to be used as an indicator. In acidic conditions, curcumin is yellow and contains two carbonyl groups in the centre of the molecule. When added to alkaline conditions above pH 8, the colour changes to red as one of the carbonyl groups changes to an alcohol group. Using a substance such as this helps students realise that chemicals are not just in bottles in the laboratory but are contained in all the substances around us.

iv) Health benefits – critical thinking

Turmeric, and its active ingredient curcumin, have been associated with a number of health benefits such as anti-inflammatory, anti-oxidant and anti-cancer properties. Turmeric supplements are widely marketed and this provides an opportunity to discuss the evidence to support claims made and the efficacy of the supplements.

3.3.3 Example 2 – Garlic

i) Introduction

Freshly chopped garlic is passed around to smell and possibly taste (disposable spoons). Students are asked to describe the smell. At the same time the structure of allicin (a key chemical constituent) is displayed (Figure 3.4). Allicin is the chemical produced that gives the chopped garlic its distinctive smell. When fresh garlic is chopped the enzyme allinase converts alliin to allicin. The chemical structure is also displayed and the students are asked to identify key components.

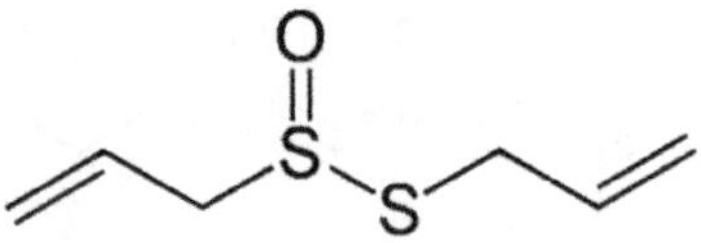

Figure 3.4 The chemical structure of allicin.

ii) Olfactory titration

Olfactory titrations have been described by a number of authors using garlic, onions and vanillin (Flair and Setzer, 1990; Wood and Eddy, 1996; Neppel *et al.*, 2005) as alternatives to detection by colour change. This has the benefit of engaging visually impaired students, and all students, with the use of a different sense to vision. The characteristic odour of garlic has proved successful for this technique. The aroma is quenched in strong basic solutions and is then released in neutral or acidic solutions. Comparing the accuracy of results from olfactory titrations to visual titrations provides opportunities to discuss accuracy and precision.

3.4 Gustatory Learning

The senses of smell and taste are closely linked and the taste of specific foods can also elicit strong memories and emotional responses. Many chemistry teachers will recognise and be aware of chemistry in food such as the Maillard reaction, jelly and ice cream or chocolate. These examples are explored in excellent resources such as Brunning (2016). However, demonstrations and explorations of these areas tend to be the preserve of outreach events or science clubs and do not appear in mainstream teaching as often as they should. Some would argue that this is because it is not specifically in the curriculum. When this material is presented, the curriculum links are not made explicit enough and the relevance is lost. For example, a lesson involving ice cream may be dismissed because it is not on the curriculum. However, the lesson may actually focus on changing states, bonding and structure and ice cream is an exemplar. Other barriers to engaging in these sorts of activities can be health and safety concerns relating to providing students with food to taste. This, however, is managed successfully by the food technology teacher because they are experienced in the procedures to safely undertake these activities. Equally, the chemistry teacher is experienced at undertaking a wide range of experiments. The food technology teacher may not be confident doing chemistry experiments just as the chemistry teacher may not be confident in undertaking food related activities. In addition, food related examples may be less familiar to the chemistry teacher. The creative teacher recognises the value of providing these opportunities to their students and thinks creatively how these activities can be linked to the specific curriculum. Thereby, it becomes a core learning activity and has curriculum time attached to it. Try identifying a challenging area of the curriculum and, using divergent thinking, see whether you can generate a range of gustatory opportunities that could be linked to this area – this could be directly or metaphorically.

Case Study 3.2 – Edible Experiments

Dr Joanna Buckley is the Royal Society of Chemistry Regional Coordinator for North East England, based in the Department of Chemistry, Sheffield University.

> *"Food is a universal medium which connects varied audiences and allows the sharing of knowledge long after the event. Once you know a fascinating fact about food chemistry, every time you revisit that food, you will remember the underlying chemistry and hopefully share this understanding with others."* (Jo Buckley).

Buckley (2019) has produced an excellent range of resources exploring the chemistry behind food. For example, a simple experiment involves a student cleaning their teeth with toothpaste and then having a drink of orange juice. The taste of the orange juice is more bitter than usual due to the presence of sodium lauryl sulfate. This compound interacts with taste receptors on the tongue, suppressing sweet receptors and enhancing bitter taste receptors. This knowledge is unlikely to be a requirement of a standard chemistry syllabus and some would therefore question the value of such an activity. However, once we explore the reason why sodium lauryl sulfate is present in toothpaste (and many other products) as a surfactant then we can link to more core chemistry knowledge such as the structure of water molecules, bonding and intermolecular forces. The initial activity acts as a hook, contextualised in an experience the student may be familiar with and provides a different sensory experience. Thinking creatively, it is possible to continue the theme of surfactants to link to bubble formation and the potential for multisensory learning that bubbles provide.

3.5 Haptic Learning

Haptic learning refers to learning emphasising the sense of touch. In chemistry teaching there are situations where this sense is engaged routinely. These include handling materials to experience how hard or soft they are, touching a test tube during an exothermic or endothermic reaction or handling materials to build models of atoms and molecules. For example, popular molecular modelling kits can be used to challenge the students to describe a molecule by touch alone. Students close their eyes (or wear a blindfold) and describe the structure to a partner who then tries to draw the structure. These kits typically distinguish between atoms of different elements by colour but this is not available to the students in this instance. They, therefore, have to come up with a different way of identifying the atoms of different elements such as by the number of bonds that the atom has (*e.g.* carbon – 4 and oxygen – 2). This has the benefit of reinforcing the bonding in organic molecules based on the number of bonds an atom can make. The students are challenged to create a diagram of the structure based on an oral description rather than observing formulae. Focusing on haptic learning can provide new challenges for the students and provides a mechanism to reinforce knowledge and understanding. In a related sense, Case Study 3.3 describes the use of manual signs (sign language) to improve understanding of the language of chemistry.

Case Study 3.3 – Manual Signs in Chemistry Learning

An example of a creative technique to improve understanding of the language of chemistry is to incorporate sign language into teaching. Jo Haywood (Parkside Federation Academy, Cambridge, UK) and Sarah Barret (Chesterton Community College, UK) undertook two small-scale studies using manual signs in teaching and learning of Chemistry.

The first of these was with year 12 students who used a combination of British Sign Language (BSL) signs (making use of the Scottish Sensory Centre glossary (SSC, 2019)) and student designed manual signs to define key terms in the topic of energetics.

They worked on terms such as ionisation energy and electron affinity (which do not have their own British Sign Language (BSL) definitions). For the new definitions, students included BSL signs for other chemistry words such as: electron, reaction, gas and temperature whilst adding in their own for mole and standard conditions.

The process of individually designing and then communicating their own manual signs meant that students thought more deeply about what the definitions meant, rather than just undertaking rote learning. This in turn, led to an increase in attainment in both straightforward definition tests and longer answer questions that required conceptual application. Students reported that they enjoyed the process of using signs despite a slight hesitance at first, and those with English as an additional language found this removed a barrier in terms of their parity of access.

The second study looked at using accepted BSL signs (Smith and Ingle, 2017) in the teaching of properties of metals. Students reported that they enjoyed using the signs, and there was increased engagement during the lesson sections where signs were used. There was found to be increased retention of knowledge where the BSL sign had been used compared to when they were not.

3.6 Conclusion

This chapter has explored using multisensory approaches to teach with creativity in chemistry. We have demonstrated macroscopic strategies that can be used to engage students with abstract chemistry sub-microscopic concepts. We have also illustrated how multisensory approaches can be embedded in a variety of contexts to engage students and reinforce chemistry understandings. While these strategies may involve more forethought, organisation and time they can help a more diverse group of students find relevance with chemistry.

References

Beghetto R. A., (2017), Creativity in teaching, *The Cambridge Handbook of Creativity Across Domains*, pp. 549–564.

Brunning A., (2016), *Why Does Asparagus Make Your Pee Smell?: Fascinating Food Trivia Explained with Science*, Ulysses Press.

Buckley, (2019), https://www.sheffield.ac.uk/chemistry/edibleexperiments.

Compound Interest, (2019), Chemical compounds in herbs and spices, https://www.compoundchem.com/2014/03/13/chemical-compounds-in-herbs-spices/.

Cox J. Rees S. W. and Banks P., (2016), Chemistry Stinks, Education in Chemistry, Available at: https://edu.rsc.org/feature/chemistry-stinks/2500085.article.

Flair M. N. and Setzer W. N., (1990), An olfactory indicator for acid-base titrations: A laboratory technique for the visually impaired, *J. Chem. Educ.*, **67**(9), 795.

Johnstone A. H., (1991), Why is science difficult to learn? Things are seldom what they seem, *J. Comput. Assist. Lear.*, 7(2), 75–83.

Lewis G. N., (1916), The atom and the molecule, *J. Am. Chem. Soc.*, **38**(4), 762–785.

Mahaffy P., (2006), Moving chemistry education into 3D: A tetrahedral metaphor for understanding chemistry. Union Carbide Award for Chemical Education, *J. Chem. Educ.*, **83**(1), 49.

Neppel K., Oliver-Hoyo M. T., Queen C. and Reed N., (2005), A Closer Look at Acid–Base Olfactory Titrations, *J. Chem. Educ.*, **82**(4), 607.

Perry N. S. L., Menzies R., Hodgson F., Wedgewood P., Howes M. J., Brooker H. J., Wesnes K. A. and Perry E. K., (2018), A randomised double-blind placebo-controlled pilot trial of a combined extract of sage, rosemary and melissa, traditional herbal medicines, on the enhancement of memory in normal healthy subjects, including influence of age, *Phytomedicine*, **39**, 42–48.

Scottish Sensory Centre (SSC), (2019), British Sign Language Glossaries of Curriculum Terms, Available at: http://www.ssc.education.ed.ac.uk/BSL/list.html.

Shulman L., (1987), Knowledge and teaching: Foundations of the new reform, *Harv. Educ. Rev.*, **57**(1), 1–23.

Smith C., and Ingle C. (2017). Let's sign science BSL Vocabulary for Key Stage 1,2 and 3. *Co-sign communications*.

Taber K. S., (2014), *Student Thinking and Learning in Science: Perspectives on the Nature and Development of Learners' Ideas*, Routledge.

Torrance E. P., (1988), The nature of creativity as manifest in its testing, *The Nature of Creativity*, 43–75.

Wood J. T. and Eddy R. M., (1996), Olfactory titration, *J. Chem. Educ.*, **73**(3), 257.

CHAPTER 4

Cultural Chemistry

Creative teaching involves developing ideas that are both novel and fit for purpose. In this chapter, we discuss the use of creative teaching strategies to engage with chemistry in different cultural contexts. It will use different case studies to illustrate how chemistry can be taught creatively in different contexts and connect chemistry to students' everyday experience. This approach has the potential to engage a broader range of students in the subject. Mahaffy (2006), who added an extra apex to Johnstone's triangle to form a tetrahedron incorporating the human element (Figure 4.1), highlighted the importance of incorporating this aspect. He argued that chemistry education benefits from a balanced emphasis on all four vertices in the tetrahedron. It helps students overcome their fear of chemistry, make imaginative connections between chemistry and their everyday lives, and provides motivation for deeper understanding. In the process, it helps students think widely, increasing the chances of making more remote connections between ideas – something at the core of applying chemistry.

Sjöström (2013) further differentiated the human element into three different levels of complexity that engage students in historical, sociological and cultural perspectives. Inclusion of the human element encourages creative chemistry educators to consider how chemistry can be situated culturally and personally within the society and school. However, as Talanquer (2013) suggested, chemistry teachers can have a limited ability to approach the teaching of chemistry in more meaningful and relevant ways. Creative chemistry teachers, therefore, should consider how their chemistry teaching can be culturally situated within the learning context in which they operate. Activity 4.1 is designed to encourage thinking about this potential within your learning context.

Advances in Chemistry Education Series No. 4
Creative Chemists: Strategies for Teaching and Learning
By Simon Rees and Douglas Newton

Published by the Royal Society of Chemistry, www.rsc.org

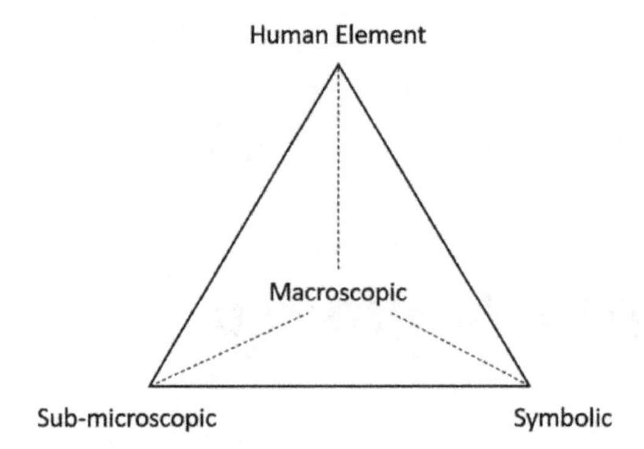

Figure 4.1 The chemistry education tetrahedron.
Adapted from Mahaffy, 2006 with permission from American Chemical Society, Copyright 2006.

Activity 4.1 – Culturally Situated Chemistry

Applying associative thinking, use the following questions to consider how to link chemistry culturally in your teaching context.

i) What are significant and relevant cultural contexts for your students?
   Think about the specific cultural context you operate in. What are important aspects of social and cultural identity in your community such as important events (past and present), significant regional employers or sporting associations?
ii) How do these contexts link to the chemistry curriculum?
   Just because the chemistry curriculum does not specifically state knowledge of chemistry in certain contexts, it does not mean that learning cannot be situated within those contexts when pedagogically advantageous.
iii) How can you incorporate these links within your teaching?
   Think about the most appropriate opportunities within the teaching schedule to incorporate this culturally situated content.
iv) What cultural resources are available in my area?
   Are there museums, theatres, sports facilities or cultural organisations in your area that could strengthen these links?

4.1 Experiential Learning

Experiential learning can engage students in authentic learning experiences in novel contexts. Cultural resources such as museums, historical sites and other visitor attractions provide opportunities to engage with the human element of the chemistry tetrahedron and explore the interactions between science and society (Hodson, 2008). In this section, we illustrate putting this into practice with two examples of cultural resources that the authors use.

Case Study 4.1 – The Oriental Museum, Durham University

Established in the 1950s to support the work of the School of Oriental Studies, the Oriental Museum at Durham University attracts thousands of visitors every year. Schoolchildren and older students visit to study the collections with a focus on humanities based subjects. However, the collection also presents an excellent opportunity to engage in multisensory learning and explore the materials that the objects are made from, their properties and how this is explained by their chemistry. This is linked to the historical and cultural stories relating to the object, illustrated by the following examples:

i) Katana or Samurai sword

Regarded by many as the ultimate sword, the katana (Figure 4.2) has been the dominant and revered weapon in Japan for over a thousand years and gave rise to a completely new art form and spiritual way of life. The Samurai blade is allegedly strong enough to cut a person in two yet also fine enough to slice a human hair.

Born in an iron smelting furnace or Tatara, the iron ore is particularly low in impurities such as sulfur and phosphorous which can make iron and steel brittle. After three days of intense heating, the molten iron combines with carbon to form steel of the correct composition for the katana. The quality of the steel was essential for establishing the standard of perfection that was to last for centuries.

The raw steel is passed on to the forge and the swordsmith who, through years of training and tradition passed down the generations, selects suitable pieces of the metal. It will take three men, three months to forge a new katana. The steel is repeatedly folded and beaten to form a steel alloy with a uniform 0.7% carbon content. Steel of a higher carbon content is wrapped around the core of the sword, a softer, more flexible steel. This prevents the sword from breaking when struck, while the outer edge is harder and very sharp. By observing the sword carefully, this area is noticeable along the length of the blade as an area that does not take on the fully mirrored appearance characteristic of the rest of the blade and is referred to as the "hamon".

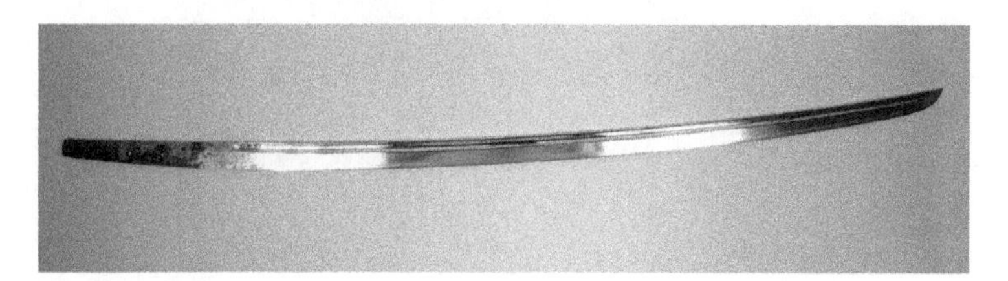

Figure 4.2 Katana (Edo Period, 1603–1868) (image courtesy of the Oriental Museum, Durham University).

This is where the softer steel core meets the harder blade and is formed by a layer of clay and carbon that is applied to the surface of the sword during the hardening process. The blade is heated to 800 °C and then rapidly cooled in water or quenched. The edge of the blade that is covered in a thin layer of clay cools more quickly than the main body and becomes harder and the blade assumes its iconic shape. The katana is polished and sharpened using seven different grades of limestone and sandstone to reveal the microstructure of the blade.

This story of the katana's creation can be linked to several areas of core curriculum content such as the reactivity series, metallic bonding and alloys. Students can make models to represent the structure at the sub-microscopic level and link this to the macroscopic and human element levels.

ii) Jade bowl

A delicate pale green jade bowl carved in shallow relief with very thin, translucent walls (Figure 4.3).

The bowl is made from the semiprecious mineral jadeite; a silicate-based mineral with sodium and aluminium ($NaAlSi_2O_6$). Belonging to the pyroxene group of minerals, the name derives from the Greek pyros (fire) and xenos (stranger). So-called because these minerals were first found in igneous deposits where they were thought to be impurities.

The structure of jadeite (see virtual museums (2019a) for a 3-dimensional representation) provides an excellent opportunity to talk about the shapes of molecules and giant structures in a new context. Within the crystal structure, silicon forms bonds with four oxygen atoms in a tetrahedral arrangement. These units form chains of infinite length with the sodium and aluminium cations between the chains. The strength of the attractions between the interweaving chains results in jadeite being an extremely hard and durable material. These "everlasting" properties have led to ground jade being consumed as an elixir thought to bestow immortality.

Figure 4.3 Jade bowl (Qing Dynasty, 1700–1899) (image courtesy of the Oriental Museum, Durham University).

iii) Earthenware

Aside from its Oriental collection, the museum also has a substantial collection of Egyptian antiquities such as 4000-year-old earthenware pots. The development of earthenware pots represented a significant stage of human development with the ability to store goods such as grain and liquids such as water, milk, oil and beer (Figure 4.4).

Kaolinite is the most important mineral in clay and consists of sheets of tetrahedral silicon dioxide (SiO_2) and octahedral aluminium oxide (Al_2O_3) which are linked through bridging oxygen atoms (see virtual museums (2019b) for a 3-dimensional representation). On the aluminium sheets, some oxygen atoms are bonded to hydrogen. These produce hydrogen bonds linking the crystal structure together. When a pot is formed, clay is mixed with some water so that it can be moulded into shape. The water molecules link the layers via hydrogen bonds that enable the clay sheets to pass over each other when guided by the potter's hands.

As the clay air dries, water molecules are lost from between the clay sheets and they move closer together. Hydroxyl groups in the kaolinite sheets then form hydrogen bonds and a stronger structure is formed. This provides an excellent example for the students to consider the role of intermolecular forces in an entirely new context. At this stage, if water was added once again the pot would disintegrate. To prevent this, the pot is heated to 1000 °C to drive off some of the chemically bound water and shorter oxygen bridges are formed (Breuer, 2012). The kaolinite loses its sheet structure and a metakaolinite 3-dimensional network forms. These chemical changes are irreversible.

Figure 4.4 Earthenware pot (12th Dynasty, 1985–1795) (image courtesy of the Oriental Museum, Durham University).

iv) Prints

Artworks can provide an excellent opportunity to engage students. In this instance, two different prints are compared. The first is a woodblock print in colour depicting the kabuki actor, Ichikawa Danjuro VII, as Matsuomaru (by Utagawa Kunisada) and the second is "Illustration of a Military Review" by Hashimoto Chikanobu showing the Meiji Emperor inspecting troops (Figure 4.5).

Printed on traditional Japanese paper or "washi", the paper is made from the inner bark of the paper mulberry bush. The fibres are boiled with sodium hydroxide solution to remove impurities such as starch, fat and tannin and then washed in ice cold water to prevent bacterial decomposition. The pulp is then spread on a screen and shaken to spread the fibres evenly. The fibres consist of cellulose; a natural condensation polymer formed from glucose molecules.

The striking difference between the two prints is the use of colour in the second print. This is due to the incorporation of Prussian Blue pigment into the painting. Developed in the West and brought to the East, it had a profound impact on Japanese art. The first modern synthetic pigment, Prussian Blue has at the heart of its chemical structure the octahedral hexacyanoferrate complex ion $[Fe(CN)_6]^{4-}$. Three of these ions are bound to four Fe^{3+} ions producing iron (III) hexacyanoferrate (II) $Fe_4[Fe(CN)_6]_3$. Prussian Blue also has therapeutic uses for treating heavy metal poisoning due to its ion exchange properties.

As these four examples demonstrate, the collection provides extensive opportunities to explore many different areas of the curriculum and find interesting links to the human element of chemistry. This engages a more diverse group of students who may have previously struggled to find relevance in chemistry to their everyday lives. For example, on one occasion, a disengaged student was able to find relevance and interest in the subject through this activity because of

Figure 4.5 Japanese prints – Matsuomaru by Utagawa Kunisada, (Edo Period, 1831) and "Illustration of a Military Review" by Hashimoto Chikanobu (images courtesy of the Oriental Museum, Durham University).

his interest in martial arts. Museums are widespread resources for this kind of experience and Activity 4.2 is designed to help you think about ways you can develop similar opportunities within your teaching.

Activity 4.2 – Chemistry Excursions

Cost and time constraints are often objections made to organising visits to different cultural resources. In some subjects, such as geography or history, however, they are regarded as essential to engage students and contextualise learning. Should this not also be the case for chemistry?

i) What excursions or field trips currently happen in your institution?
   Rather than organising a new trip somewhere, think about the trips that are currently organised for students and build chemistry learning experiences into them.
   SCAMPER (Chapter 2) may be a useful approach to think about this.
ii) How can chemistry learning be incorporated into these trips?
   Thinking divergently, how can these trips offer the potential to make chemistry engaging? For example, a trip to a Roman site could consider the materials used to make different artefacts.
iii) Is there a different excursion that could be developed?
   Using creative thinking strategies described in Chapter 2, think about developing a new opportunity that incorporates chemistry as well as other subjects.

Case Study 4.2 – Beamish Museum

Beamish museum is a large, open-air "living" museum in the North East of England. Visitors explore different areas of the town covering periods from the 17th to 19th century. Visited by tens of thousands of people and students of all ages, it enables them to explore stories of social history. There are varied opportunities to connect to the chemistry curriculum, illustrated by the following examples:

i) The Victorian pharmacy
   The shift from rural to urban living conditions in Victorian Britain gave rise to poor health. At this time, a visit to the doctor was a luxury for only the wealthy. The pharmacist was the main source of medical provision and gave birth to the High Street chemist that we would recognise today (Eastoe, 2010). The Victorian pharmacy contained a vast array of patented medicines, concoctions and remedies supposedly able to treat every conceivable illness. Many of these would now be considered bizarre, dangerous or illegal with ingredients such as arsenic, asbestos, opium, cocaine and cannabis.

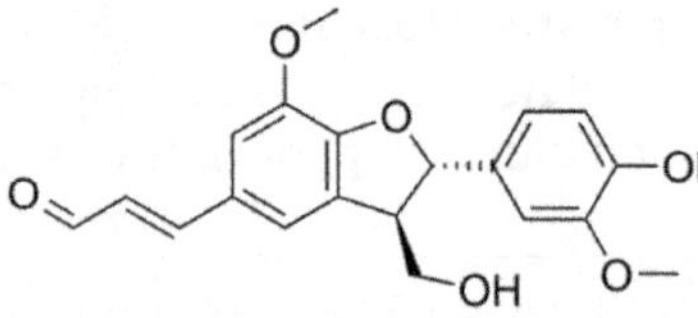

Figure 4.6 Structure of balanophonin, a chemical constituent of Sang Draconis.

A visit to a pharmacy, therefore, affords excellent opportunities to explore the chemical constituents of different concoctions. For example, Sang Draconis (Dragon's blood) is a bright red resin extracted from the dragon palm (*Dracaena draco*). Dragon's blood has been used since Roman times as a medicine to cure fevers, treat intestinal problems and diarrhoea. The chemical structure of one component of dragon's blood, balanophonin is shown in Figure 4.6. The context can be used to teach organic chemistry structures and reactions.

Essex (2013), working with a group of students, analysed the contents of bottles from a Victorian pharmacy at Ironbridge Gorge in England. Inorganic chemical tests for halides and sulfates and flame tests for cations were undertaken as well as analytical techniques such as infrared spectroscopy and nuclear magnetic resonance to analyse organic samples. Students engaged in a range of relevant inorganic chemistry content in a completely new context.

To recreate this activity, we have produced our own bottles from a fictitious Victorian pharmacy (Figure 4.7). and students devise and undertake a range of chemical tests to investigate the contents. They consider the chemical structure of the possible contents and then select appropriate tests. The labels may give some indication of their contents but some of the bottles do not contain what is expected.

Figure 4.7 Recreated Victorian pharmacy bottles.

For example, the bottle labelled "glacial ethanol" does not give a positive result for a primary alcohol but does fizz with magnesium; suggesting the ethanol has oxidised to ethanoic acid over time.

ii) Sweet shop

Across the road from the pharmacy is the sweet shop where a range of traditional sweets are produced. Victorian confectioners produced many classic sweets such as pear drops, fruit pastilles and sherbet lemons that remain popular today. The industrial revolution brought about many technological advances and falls in the price of refined sugar that enabled factories to produce confectionery that many people could afford. There are simple recipes for making sweets (Husband 2014; bbcgoodfood, 2019) which can be linked to a range of areas of the chemistry curriculum such as: states of matter, bonding and structure, flavourings, colourings.

For example, most sweets are produced by dissolving sucrose in water. Exploring this process promotes discussions about the process of dissolving (and how this is different to reacting), bonding, structure and intermolecular forces.

Husband (2014) also explains how this process can be used as an example of a dynamic equilibrium (a challenging concept for many students) as the point of saturation is reached and the rate of dissolving is equal to the rate of crystallisation. The sweet shop also provides opportunities to link to multisensory chemistry (see Chapter 3) and discuss health issues such as sugar consumption and artificial sweeteners.

iii) Coal mine

Many parts of Britain lie on strata of coal that led to coal mining and a strong mining heritage. Culturally, coal mining remains a significant part of the regional identities of these areas. Beamish museum provides the opportunity for visitors to walk down a shallow shaft and experience the working conditions of the miners.

Climate change is a significant issue for all and has increasingly galvanised young people into action (BBC, 2019a). This provides the opportunity to discuss the environmental impact of coal burning and the shift in energy production away from coal fired power stations (BBC, 2019b). With creative thinking, there is a variety of ways this social context can be linked to the chemistry curriculum.

Calculating energy changes associated with different fuels is a common activity that can also be applied to comparisons with coal and natural gas (Activity 4.3) and the carbon dioxide released during this process. Using chemistry knowledge in different contexts enables students to develop their associative thinking skills and gain confidence in their abilities. The environmental issues can be taken further by considering other components of coal such as sulfur content and the consequent production of sulfur dioxide and acid rain.

Activity 4.3 – Comparing Coal and Natural Gas Combustion

UK energy production has shifted from coal to a mix of renewables (e.g. wind, solar, biomass), nuclear and natural gas. Electricity production from natural gas is claimed to produce half the carbon dioxide compared with coal burning. This activity demonstrates how students can apply knowledge from the chemistry curriculum to investigate this claim.

i) Students are asked to derive the chemical equation for the combustion of natural gas (assumed to consist of methane).

$$CH_4\ (g) + O_2\ (g) \rightarrow CO_2\ (g) + H_2O\ (g)$$

ii) Using bond enthalpy data, the students determine the enthalpy change for the combustion of methane as around $-690\ kJ\ mol^{-1}$

This is significantly lower than the published value of $-888\ kJ\ mol^{-1}$ and can promote creative thinking as to why this is the case.

iii) Coal consists mainly of carbon and combustion is represented by the following equation:

$$C\ (s) + O_2\ (g) \rightarrow CO_2\ (g)$$

The enthalpy of combustion for this reaction is $-393\ kJ\ mol^{-1}$.

Therefore, approximately double the amount of coal would be required to be burned to release a similar amount of energy and double the amount of CO_2 would be produced.

This data can be used in different ways to reinforce chemistry learning. For example, if a coal fired power station burns 1.5 million tons of coal per year, how much carbon dioxide is produced per year?

4.2 Cultural Chemistry in the Classroom

Another approach is to think creatively about how chemistry can be culturally contextualised in the classroom. For example, Kristy Turner (Bolton School Boy's Division) uses an article (Chapman, 2017) describing an arsenic poisoning case that occurred in Bradford in 1858. Over 200 people were poisoned and 21 people died when sweets were mistakenly made containing arsenic trioxide. Year 8 pupils (age 12–13) recount it in their own words as a newspaper article. The article is also available as a podcast, which made it more accessible to students who are unfamiliar with some of the technical words.

The context of the story is important. The poisoning case occurred in Bradford, a mill town in Northern England very similar to Bolton. This makes the story meaningful to students in their cultural context. In preparation for the assignment, the pupils look at pictures of both towns at the time, especially pictures of shops, aspects of the characters' lives, how well educated they were and how this contributed to the events that unfolded.

Case Study 4.3 – Creative Chemists

Kristy Turner

(Image courtesy of Kristy Turner)

Kristy teaches chemistry in a unique hybrid role across the school-university transition. She teaches chemistry in a secondary school (Bolton School Boy's Division) and at the University of Manchester, UK.

What is your background that led you to this point?

Following my PhD and a short period in the pharmaceutical industry I trained as a secondary chemistry teacher. I thrived in this job and was quickly promoted to Head of Department. In 2011–12, disillusioned with middle management and seeking to reconnect with my subject, I took a sabbatical and spent a year in the School of Chemistry at the University of Manchester as a school teacher fellow. Following this I went back into teaching as a mainscale teacher of chemistry and biology with no managerial responsibility. I am particularly interested in the transition between school and university for students in chemistry and in 2015 I designed a new role for myself, combining school teaching and lecturing and education research in chemistry at the University of Manchester.

How has creative thinking been important in your career?

Creative thinking has been a key driver in my career. My role is unique in the UK and I essentially designed the role myself. Through creatively exploring the issues faced by HE and schools I have been able to design a role that bridges the transition and shares expertise across both sectors. My research benefits from me approaching chemistry education issues through different lenses to design creative solutions.

Jane Essex (Strathclyde University) uses the kelp industry to incorporate chemistry within a relevant cultural context in Scotland. The separation and purification of substances is a central theme in chemistry and occurs at all levels of the secondary school chemistry curriculum. Scottish policy encourages teachers to consider the distinctive cultural context of the country. Against this backdrop, student teachers considered the role of seaweed harvesting and roasting ('kelping') in the country's past. They considered how the distinctive chemistry of seaweed, including its high concentrations of carbonate (around 5%) and iodide ions (between 1% and 6%), has culinary value. This, in turn, raises the possibility of harvesting seaweed commercially for consumption.

Crofters created 'lazy beds' *(feannagan)*, of seaweed alternating with peat sods. These were ideal for growing potatoes, a staple food item for the crofters. In addition, the small sheep they raised were able to climb down to the beaches to graze on seaweed.

An alternative use was as a cash crop, selling the seaweed ash for use in glass making. This became increasingly important during the Napoleonic wars, when European plant ash sources of sodium carbonate were unavailable. The wars ended and alkali production from salt began soon afterwards, causing the collapse of the kelp industry. Meanwhile crofters who had been forced by estate owners to carry out kelping had neglected their land, because peak kelping time coincided with the busiest period of the agricultural year. Moreover, the estates had been populated with large efficient sheep breeds (*caora mor*) that were unable to climb down to the beaches. Finally, crofters had not used seaweed to prepare their lazy beds, ready for future crops. These factors cumulatively contributed to the widespread abandonment of the crofts. This historical backdrop provides a context in which students go on to extract iodine from seaweed.

Although kelping exploited the high sodium carbonate concentrations in the seaweed, iodide is similarly concentrated. The extraction of iodine from laminaria seaweed is described by the Royal Society of Chemistry (RSC, 2019). The production of the ash is a lengthy process and smoky, so needs to be done in fume cupboards which may be scarce in schools. This limitation can be linked to the amount of time it would have taken the crofters away from their land. This problem can be overcome in the laboratory by having a technician pre-burn the seaweed. The iodide is then dissolved in hot water, filtered and the iodide ions are oxidised to produce iodine solution. The amount of residue left at this stage points to the resource inefficiency of taking just one component out, rather than using it all. Finally, the iodine is extracted using a suitable organic solvent, in which the iodine preferentially dissolves. The solvent is left to evaporate and iodine crystals form.

In undertaking experiments in this context, the creative chemistry teacher is demonstrating to the students the challenges and rewards of obtaining useful substances from natural substances. It also highlights the ingenuity of their ancestors and the importance of chemistry to enable them to survive in challenging conditions.

4.3 Conclusion

In this chapter, we have demonstrated, with a variety of case studies, how chemistry can be made more relevant to a greater diversity of students by situating the chemistry learning within culturally relevant situations. Opportunities vary with location but using it imaginatively can bring relevance and personal meaning to chemistry that engages students of all ages and integrates their learning across disciplines.

References

BBC, (2019a), Students miss school to protest climate change, https://www.bbc.co.uk/news/av/uk-england-london-47256691/students-miss-school-to-protest-climate-change.

BBC, (2019b), UK has first coal free week for a century, https://www.bbc.co.uk/news/business-48215896.

BBC Good, Food, (2019), Peppermint creams, Available at: https://www.bbcgoodfood.com/recipes/peppermint-creams.

Breuer S., (2012), The chemistry of pottery, *Educ. Chem.*, **49**(4), 17.

Chapman K., (2017), How a 'daft' pharmacy mix up led to a series of poisonings in Victorian Britain, https://www.chemistryworld.com/podcasts/arsenic-trioxide/2500445.article.

Eastoe J., (2010), The Victorian Pharmacy.

Essex J. (2013), Behind the scenes at the Victorian Pharmacy, https://eic.rsc.org/feature/behind-the-scenes-at-the-victorian-pharmacy/2020098.article.

Hodson D., (2008), *Towards scientific literacy: A teachers' guide to the history, philosophy and sociology of science*, Brill Sense.

Husband T., (2014), The sweet science of candy making, Available at: https://www.acs.org/content/acs/en/education/resources/highschool/chemmatters/past-issues/archive-2014-2015/candymaking.html.

Mahaffy P., (2006), Moving chemistry education into 3D: A tetrahedral metaphor for understanding chemistry. Union Carbide Award for Chemical Education, *J. Chem. Educ.*, **83**(1), 49.

RSC, (2019), Extraction of iodine from seaweed, Available at: https://edu.rsc.org/resources/extracting-iodine-from-seaweed/1915.article.

Sjöström J., (2013), Towards Bildung-Oriented Chemistry Education, *Science & Education*, **22**(7), 1873–1890.

Talanquer V., (2013), Chemistry education: Ten facets to shape us, *J. Chem. Educ.*, **90**(7), 832–838.

Virtual museums, (2019a), https://virtual-museum.soils.wisc.edu/display/jadeite/.

Virtual museums, (2019b), https://virtual-museum.soils.wisc.edu/display/kaolinite/.

CHAPTER 5

Constructing and Representing Understandings in Chemistry

In this chapter, we consider the different ways that creative thinking is used to construct relationships and represent understanding in chemistry. We evaluate different representations of the Periodic Table and the use of multiple representations to develop sub-microscopic understandings. This is to illustrate how creative and critical thinking is applied in chemistry, and how this can lead to opportunities for creative thinking in the classroom – both to support the teacher's creative ideas and the students' response to them. In particular, it illustrates how ideas and models that become commonplace can be seen in a new light. This is often the source of a productive creative idea.

5.1 Creative Thinking and the Periodic Table

The essence of developing creative thinking through problem solving is to present a problem that has potentially multiple possible solutions. In chemistry, devising solutions to display the relationships between all the elements in the universe in a concise format has provided just such a challenge and has exercised considerable creative thought.

Since Dmitiri Mendeleev developed the basis of the modern Periodic Table, its form has become a familiar cornerstone of chemistry teaching. Refined into its currently accepted arrangement, it is an iconic and ubiquitous representation of the chemical world (Figure 5.1) and interactive versions are readily available (*e.g.* RSC, 2019). The Periodic Table encapsulates human understanding about the relationships between the fundamental building blocks of the universe and is a demonstration of creative thinking in chemistry. In addition, the iconography of the table has assumed

Advances in Chemistry Education Series No. 4
Creative Chemists: Strategies for Teaching and Learning
By Simon Rees and Douglas Newton

Published by the Royal Society of Chemistry, www.rsc.org

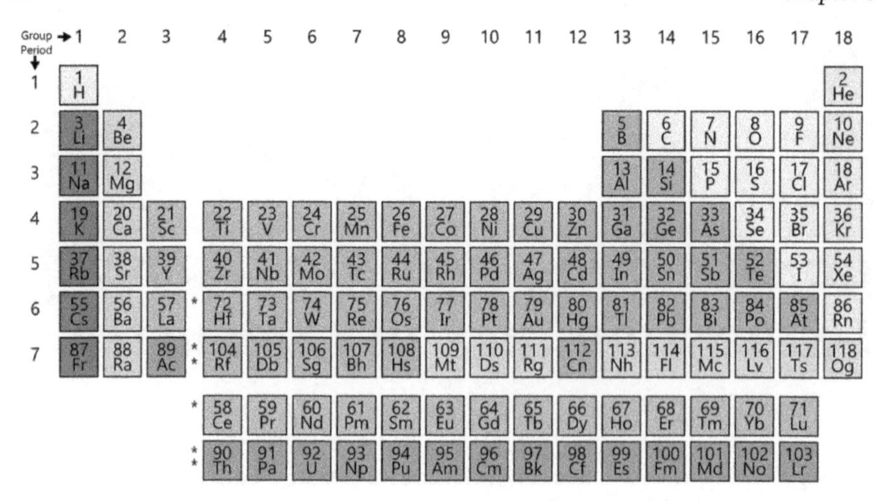

Figure 5.1 The Periodic Table.
Reproduced from https://commons.wikimedia.org/wiki/File:Simple_Periodic_Table_Chart-en.svg under the CC0 1.0 Universal Public Domain Dedication, https://creativecommons.org/publicdomain/zero/1.0/deed.en.

wider cultural significance with the symbols of the elements used creatively in different settings such as in advertising and television.

While curricula may attempt to discuss the development of the Periodic Table, this is typically limited to a historical account recalling the ordering of elements by atomic weight, recognizing repeating patterns of well-known properties. These are examples of convergent thinking, where order is applied to the known elements properties and trends and relationships identified. Mendeleev's improvements left gaps for elements not yet discovered and reordered some elements to align them by similar properties. This is an example of divergent thinking in order to address difficulties where the relationship broke down. Often, however, the Periodic Table, as with much curriculum content, is presented as a "fait accompli" and students learn to interpret the table and develop their understanding. Aside from its role as a representation of relationships, the Periodic Table affords an excellent opportunity to explore divergent and evaluative thinking in chemistry by considering different representations.

Glenn Theodore Seaborg (1912–1999) was the co-discoverer of 10 transuranium elements and – along with Yuri Oganessian – holds the honour of being one of the only two people who have had an element, Seaborgium and Oganesson, named after them in their lifetime. He made the most significant contribution to the design of the Periodic Table since Mendeleev, resulting in the version that we are familiar with today. Seaborg proposed a new row (the actinides) that contained elements similar to the "rare-earth" lanthanides and added it below (Scerri, 2011). Despite being warned that it would be professional suicide to pursue the proposal he developed the version of the Periodic Table which now adorns the walls of chemistry laboratories the world over. This persistence in the face of accepted opinion is a typical attribute of a highly creative person. Similar arrangements had, however, been proposed by Henry Bassett in 1892 (Leach, 2019a) and Alfred Werner produced virtually the modern table in 1905 (Leach, 2019b).

The elements are presented in order of atomic number and elements with similar properties appearing in vertical groups. This representation, however, has several limitations, such as the inconsistent positioning of hydrogen that contains one electron in the outer shell. This element is often positioned above lithium in Group 1 but its properties as a diatomic gaseous non-metal are very different. Sometimes, hydrogen is placed on its own or it is aligned with the halogens – Group 17 – given hydrogen's physical properties and the fact that it requires only one electron to fill its outer shell. Secondly, there is also the unnatural break at the end of each period where the reader has to jump back to the start of the next row rather than being able to count along continuously. Thirdly, there is the break in the atomic number continuum due to the separation of the lanthanides and actinides below the main table. The table is arranged in blocks corresponding to the s,p,d,f orbitals that the outermost electrons occupy. The lanthanides (atomic number 57–71) and actinides (atomic number 89–103) are the f block elements and are depicted in a separate block below the main table. This is challenging for novice chemists, because it requires an understanding of electron configurations in order to be able to understand this arrangement. The question, therefore, arises – are there alternative ways to arrange the elements?

In response, teachers, chemists and other iconoclasts have engaged in creative divergent thinking to develop hundreds of different representations of the Periodic Table. A comprehensive list of these is available at the internet database of Periodic Tables (Leach, 2019c) which contains examples of many different designs, including 3-dimensional arrangements, spirals and helices. These different representations arise from divergent thinking about how to organise relationships between the elements. Some of these different designs are depicted below to illustrate the creative thinking behind the ideas.

5.1.1 Benfey's Spiral Snail

The first graphic representation of the periodic system of the elements was a spiral wound around a cylinder, designed by the French mineralogist, Alexandre-Emilé Béguyer de Chancourtois. Mendeleev himself stated, "*In reality the series of elements is uninterrupted, and corresponds, to a certain degree, to a spiral function.*" (Jensen, 2002, p. 56). Charles Janet (1928) developed helical arrangements of the elements and is regarded by Stewart (2010) as the unrecognised genius of the periodic system. However, the 3-dimensional nature of the representation failed to gain wide appeal partly due to the difficulties of representing this in two dimensions for printing.

Two-dimensional representations of a spiral arrangement, such as Theodore Benfey's Periodic Table first published in 1964 (Figure 5.2) (see Benfey, 2009) have since been developed.

At first glance, the appearance of this table, likened to a snail, can be confusing when compared to the familiarity of the accepted arrangement. However, by focusing on hydrogen in the middle and then following the elements spiraling out from it, more sense can be made of the arrangement. Using evaluative thinking, the advantages and disadvantages of this arrangement can be considered. For example, the spiral arrangement removes interruptions in the atomic number continuum and the periodic divide is indicated at the appropriate point. Furthermore, the transition metals, lanthanides and actinides extend out from the central spiral. As the loops of the spiral increase in size further from the centre, this also provides a visual representation of increasing atomic radius with increasing atomic number. Disadvantages may include the fact that there are now no clear periods and, therefore, no clear association with energy levels or electron shells.

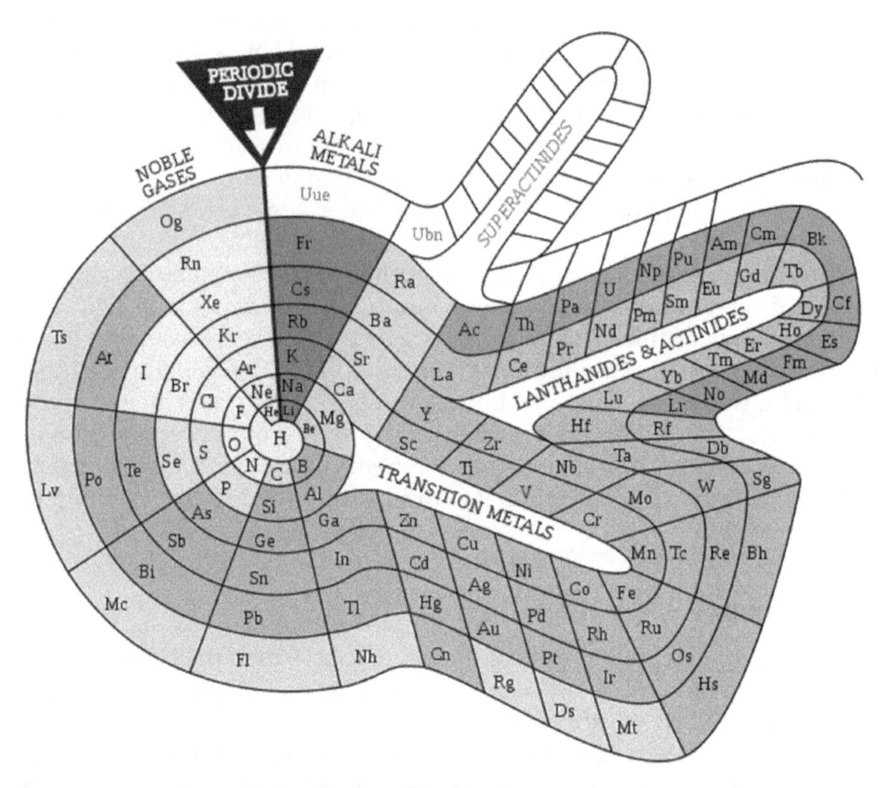

Figure 5.2 Benfey's spiral Periodic Table (DePiep, 2019). Reproduced from https://en.wikipedia.org/wiki/Alternative_periodic_tables#/media/File:Elementspiral_(polyatomic).svg under the terms of the CC BY-SA 3.0 licence, https://creativecommons.org/licenses/by-sa/3.0/.

5.1.2 Chemical Galaxy

The spiral representation has been conceived in a variety of forms including being incorporated into the concentric rings as a galaxy (Figure 5.3). In this format, the association between the elements as the building blocks of the universe is made explicit in a colourful and aesthetic way that is visually appealing, potentially more memorable and engages a wider diversity of students. As Stewart (2004) says "The objective is to show the shape of the whole and to express the beauty and reach of the periodic system".

The elliptical design enables more elements to be incorporated in the outer rings while maintaining the periodic sequence. At the very centre is a question mark representing element 0. This is the hypothesized element "neutronium" consisting of one neutron. Mendeleev himself believed that there would be an element of "group zero in period zero". Neutronium is thought to exist in neutron stars and form when nuclear fusion no longer supplies enough energy to counteract gravity. The nuclei and electrons of all the elements are crushed together to become so dense that a thimbleful would weigh 300 million tons (Chemicalgalaxy, 2019). Activity 5.1 is designed to encourage students to engage in speculative creative thought and to be comfortable with generating questions for which the answer may not be known.

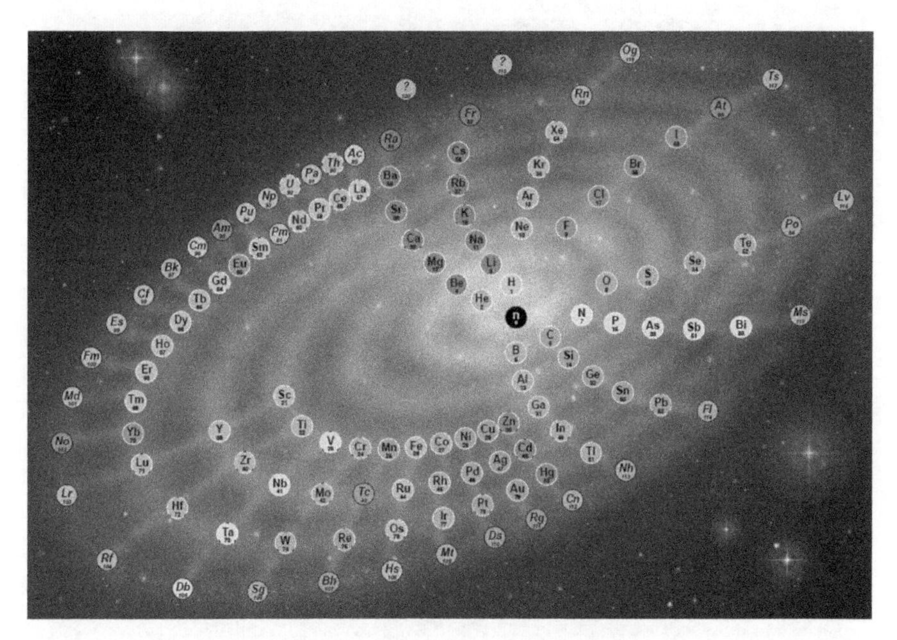

Figure 5.3 Chemical Galaxy III – A new vision of the periodic system of the elements (Image courtesy of Philip Stewart).

Activity 5.1 – Focused Imagination

The following questions are designed to encourage students to engage in speculative creative thought and explore their understanding of the nature of elements and substance.

i) What would a substance made entirely of neutronium be like?
ii) What would be its physical and chemical properties?
iii) Would it be possible to exist?
iv) How could we begin to answer these questions?

5.1.3 3-Dimensional Representations

The Periodic Table has also been reconceptualised in a variety of 3-dimensional forms such as pyramids, cubes and ribbons. For example, the Alexander Arrangement of Elements (Figure 5.4) takes the linear format of

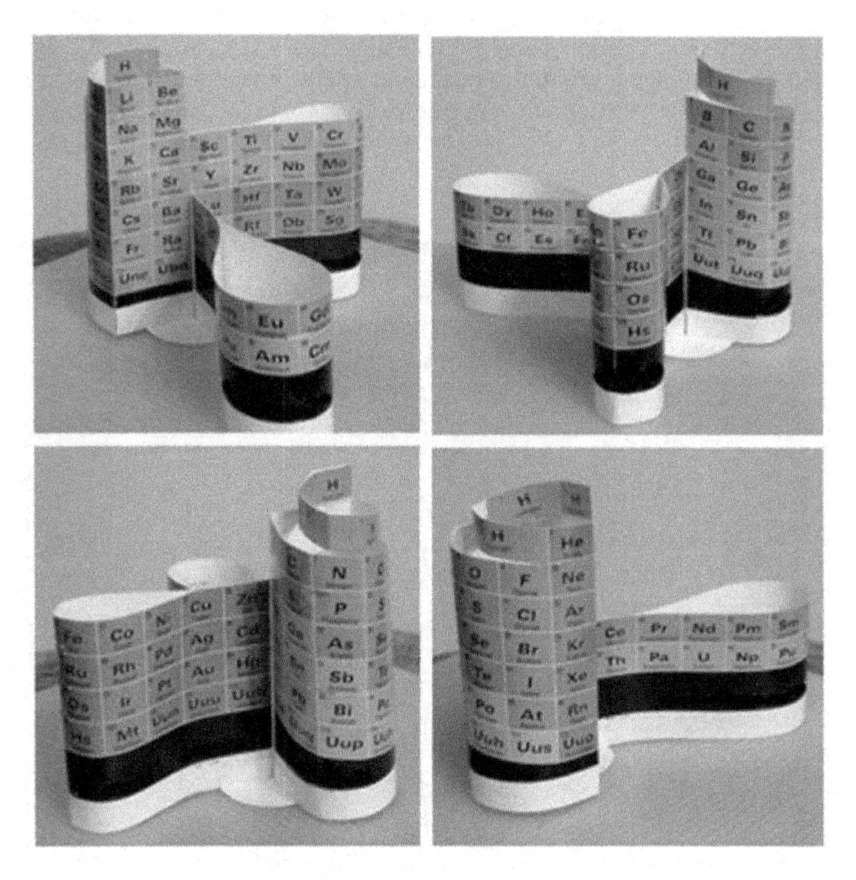

Figure 5.4 The Alexander Arrangement of Elements (Image reproduced courtesy of Roy Alexander).

the Periodic Table and turns it into a 3-dimensional arrangement with the d and f blocks looping out from the s and p blocks (Alexanderdesign, 2019). In this way, the atomic number continuum is maintained while also indicating periodicity. Seaborg (1985) admitted that the f-block had been more correctly positioned by Roy Alexander's three-dimensional element arrangement. The similarities between this arrangement and Benfey's spiral snail are apparent.

This 3-dimensional representation produces a previously unobserved aspect – a "nexus" – where three blocks of the Periodic Table start from a common location.

5.1.4 Periodic Table of Element Scarcity

Other forms of the Periodic Table can be used to highlight particular issues. For example, the European Chemical Society's Periodic Table of Scarcity (EuChemS, 2019) (Figure 5.5). The area occupied by each element is an indication of the abundance of that element and the colour coding indicates the risk of running out of the element. It illustrates how different representations of the same thing can have different purposes. This resource can be used as a stimulus to promote discussions about the exploitation of resources, the importance and challenges of reuse and recycling.

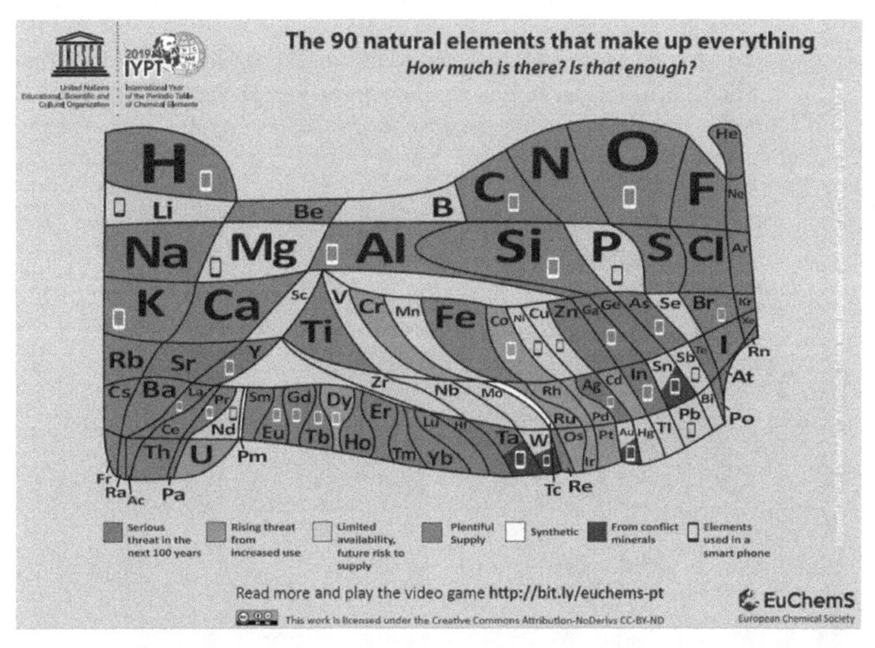

Figure 5.5 The Periodic Table of Scarcity (EuChems, 2019).
Reproduced from https://www.euchems.eu/euchems-periodic-table/ under the Creative Commons Attribution-NoDerivs CC-BY-ND, https://creativecommons.org/licenses/by-nd/3.0/.

The rich variety of different representations of the Periodic Table illustrates the capacity for human creative thought to provide different perspectives and solutions to problems. Given the technological world that we now live in and the ability to represent 3-dimensional objects on computer screens, maybe the time is now right for these representations of the Periodic Table to achieve more prominence.

Activity 5.2 aims to guide students through the process of critically evaluating the Periodic Table and then generating their own ideas for alternative arrangements. This can be an illuminating process as they may not have previously had the opportunity to reflect on this and the established arrangement was accepted without question.

Activity 5.2 – Evaluating the Periodic Table

In this series of activities, students are guided through a creative thinking process designed to explore understandings of the elements and the relationships between them.

i) Present the students with the Periodic Table and, using evaluative thinking, ask them to generate questions and observations about its organisation. Questions to consider could include:
   - What are the key features of how the elements are organised?
   - Why are they organised in this way?
   - Are there any aspects that are problematic or unclear such as the positioning of hydrogen or the separation of the f-block?

ii) Based on the observations in (i) the students engage in divergent thinking to generate alternative ways of organising the elements that may address shortcomings of the established arrangement. The students then explain their new arrangement. Figure 5.6, for example, shows two examples where students generated spiral based ideas.

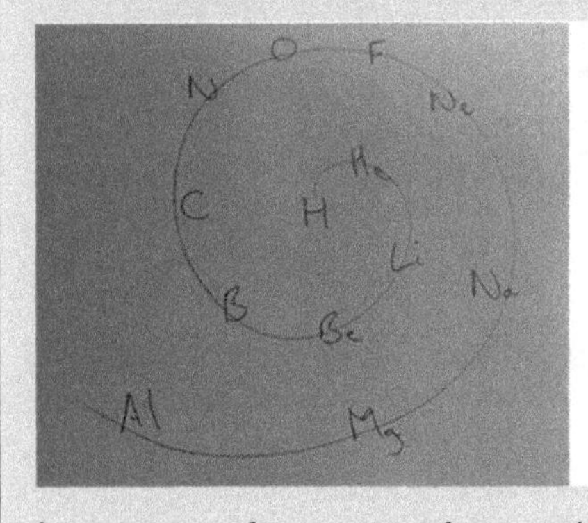

Figure 5.6 Student-generated reorganisations of the Periodic Table.

iii) The students are then presented with alternative representations such as the Chemical Galaxy to compare and evaluate in relation to their own ideas.

5.2 Multiple Representations

Conceptual understanding in chemistry requires the development of mental models that enable students to visualise the sub-microscopic, macroscopic and symbolic levels and the connections between them. A mental model is a set of mental representations and the associated mental operations that adapt and apply these to solving a problem (Merrill, 2002). Chemistry presents unique challenges in this way. From an early stage, students are required to use their imagination to develop understanding. Furthermore, multiple representations are used to represent the same phenomenon in order to emphasise different aspects (Kozma, 2003; Ainswoth, 2008; Nyachwaya and Gillaspie, 2016). Chemists use multiple representations to solve complex problems and can switch from one representation to another (Kozma and Russell, 1997).

Consider, for example, the different ways that the chemical bonding in hydrogen chloride is represented. This may be as the chemical formula, a dot and cross diagram, a line connecting the two atoms, a molecular orbital diagram or a space filling model. Each of these representations convey different information and these different representations can illicit different interpretations from students (Allred and Bretz, 2019). Multiple representations have been shown to enhance conceptual understanding (Kozma, 2003; Seufert, 2003; Ainsworth, 2006). However, for novice learners, multiple representations can cause difficulties and increase cognitive load (Corradi *et al.*, 2014; Nyachwaya and Gillaspie, 2016). Therefore, the creative teacher must decide on the most effective use of representations in their teaching context.

Activity 5.3 – Multiple Representations

This activity encourages the development of evaluative thinking by consideration of different representations.

i) Present the students with a range of representations for a chemical compound such as sodium chloride. These may include:
   - chemical formula
   - dot and cross diagram
   - ball and stick models (2D and 3D)
   - space filling models

ii) Ask the students to reflect on the information that different representations provide.

iii) Ask the students to reflect on the limitations of each representation and the misconceptions they can convey.
iv) Ask the students to design their own model to convey their understanding of the compound.

5.3 Visual Literacy and Spatial Ability

Visual literacy involves the ability to understand, produce, and use images, objects, and visible actions (Debes, 1969). It recognises the importance of images in conveying meaning. Imagery is becoming increasingly dominant in the internet age as photographs and illustrations are ubiquitous compared to more traditional printed formats. Some argue that this is creating a generation that are intuitive visual communicators and are more visually literate than previous generations (Oblinger and Oblinger, 2005). However, exposure to a wide range of imagery does not necessarily lead to a greater degree of visual literacy in interpreting and evaluating the images. With training, students can develop their abilities to interpret imagery.

Visual literacy and spatial ability are closely linked and spatial ability is correlated with success in chemistry (Coleman and Gotch, 1998). Harshman *et al.*, (2013) illustrate this point with the example of a 2-dimensional wedge and dashed representation of methane. A student with visual literacy, but lacking spatial literacy will be able to interpret the wedges and dashes as indicating bonds protruding in front and behind the plane of the paper but will not be able to imagine a 3-dimensional model of the molecule. In contrast, a student with limited visual literacy but enhanced spatial ability, will be able to envisage and rotate a 3-dimensional mental model of the molecule but would find it difficult to interpret the visual image. Hence, in any classroom, there will be students where this presents challenges in a variety of ways and the creative teacher will design pedagogy to assist students in developing these skills.

5.4 Visualisations

The development of on screen visualisations and animations (external representation) provides an opportunity to help students develop their mental models (internal representation). The Phet simulations (University of Colorado, 2019), for example, enable students to manipulate different experimental scenarios with the macroscopic, sub-microscopic and symbolic levels clearly represented. This helps students to develop integrated mental models across these three levels. Students are also able to click between different representations of molecules such as space filling and ball and stick. Alternatively, analogy based simulations can use objects in a familiar situation to help students understand abstract entities in a chemical system (Ashe and Yaron, 2013).

Students can have difficulty translating a 2-dimensional image of a molecule into a 3-dimensional shape. Jmol is a visualisation tool that is an open-source Java viewer and has been used to create 3-dimensional rotatable molecular models. Models are also available for biochemical molecules such as haemoglobin (Oliver-Hoyo and Babilonia-Rosa, 2017) *via* the Protein Data Bank (RCSB, 2019).

An important consideration in the design of these resources is the extent to which the model is simplified. Animations have the potential to convey a more realistic representation of the sub-microscopic level but detailed representations can be over complicated and result in cognitive overload (Sweller, 1994). Some animations enable students to control whether the animation is simple or realistic by using a slider that changes the degree of complexity. This enables the student to appreciate the overall complexity but also focus on key interactions.

Aside from 3-dimensional visualisations on a screen, physical kits and models are important resources to enable students to engage in multi-sensory learning and develop their mental models. Molecular modelling kits, for example, provide components to build a wide variety of molecules in three dimensions as well as visualise orbitals. It is also possible to build molecular models out of any sort of everyday material such as playdough, spaghetti, balloons, marshmallows or beads (Turner, 2019) (Activity 5.4).

Activity 5.4 – Molecular Models

This activity promotes creative thinking by requiring the students to use a selection of materials to build molecular models and then to evaluate them.

i) Provide the students with a range of materials such as pipe cleaners, playdough, short lengths of spaghetti, wire, springs, *etc.*
ii) The students build models of a range of molecules such as methane, glucose, diamond, polythene.
iii) The students evaluate the strengths and weaknesses of the model, *i.e.* its limitations as a representation of a real molecule.

Models can be used to help students visualise abstract scientific concepts such as pressure (Han and Kim, 2018) and the model can act as a metaphor to represent an abstract concept (Gilbert and Justi, 2016).

5.5 Role Play Representations

One of the limitations of physical models and many virtual simulations is that they do not show the dynamism, movement and energy that is fundamental to any chemical system. The different representations of a molecule of hydrogen chloride may show the bonding electrons in a variety of different ways but they do not give a sense of the vibration within in the bond, the movement of electrons or charge distribution. Engaging students in role-play to create representations of chemical phenomena can help them develop their mental models and reflect on their current understanding. Role-play can be used to build atoms, with different members of the class adopting roles as protons, neutrons and electrons. This can then be extended to explore chemical bonding and the formation of compounds. It can also be used to help students interpret the chemical equations and the symbolic level by simulating chemical reactions such as equilibrium reactions with students joining together and coming apart again or concepts such as rates of reaction.

Activity 5.5 – Role Play

This activity encourages you to think of new opportunities where you could develop representations using role play.

(i) Think of some different areas of the curriculum and what strategies you currently use to help the students develop understanding.
(ii) Is there an area that you think the students particularly struggle with?
(iii) How could you develop a role play representation activity to address this area?
(iv) What pedagogical justification is there to use role play in this instance?

5.6 Analogies and Metaphors

Analogies and metaphors are powerful language tools that use one thing to link to and represent the characteristics of another. Metaphors imply that two distinctly different things share common characteristics. They are used in creative writing in order to add emphasis, help the reader see things from a new perspective and create a more interesting and enriched image of the scene in a story.

While there are differences between using a metaphor for poetic effect and using it to enhance knowledge (Beall, 1999), this creative writing skill is analogous to the situation in chemistry teaching where analogies and metaphors are used to represent the characteristics of chemical phenomena. Analogical reasoning is an important aspect of the imaginative component of the five habits of mind for creative thinking (Lucas and Spencer, 2017). They are, therefore, a valuable tool for the creative chemistry teacher to help students develop an understanding of the abstract sub-microscopic level.

They can emphasise key points, create new perspectives and enhance students' mental models. They are designed to enhance understanding of abstract and unfamiliar concepts by linking with contexts and situations with which the student is familiar (Treagust *et al.*, 1998). The value of this technique is highlighting the commonalities that the two systems have, captures parallels across different situations (Gentner, 1998) and facilitates the transfer of relationships from the known to the unknown. Analogies have a role in restructuring students' conceptual frameworks (Duit *et al.*, 2001) and providing an analogical bridge to communicate knowledge of the topic (Treagust *et al.*, 1998). Newton (2012) summarises this process as shown in Figure 5.7.

A better-known parallel is recalled and related to what is to be understood, this parallel (the analogue) is articulated or run to produce the parallel effect, and the parallel effect is translated into the target situation. This avoids the more difficult route indicated by the deleted arrow. It is not unusual for creative thought in science to be supported by analogical reasoning, both amongst professional scientists and when working with students (see Newton, 2012).

For example, to say an atom is like the solar system can convey appropriate mental models linking the sun and the nucleus of the atom and planets as electrons circulating around the nucleus. However, analogies are not always necessary or appropriate and it is important to be aware of both the power and the pitfalls of analogies (Duit *et al.*, 2001). If the metaphor is extended too far, misconceptions (Heywood and Parker, 1997) and incorrect mental models can be developed such as electrons as spheres circling around the nucleus in specific orbits. Beall (1999) argues, therefore, that a metaphor can be counterproductive when there are aspects of the metaphor that do no enhance understanding of the phenomena. A metaphor can be taken too literally and lead to misunderstandings. There may also be problems with

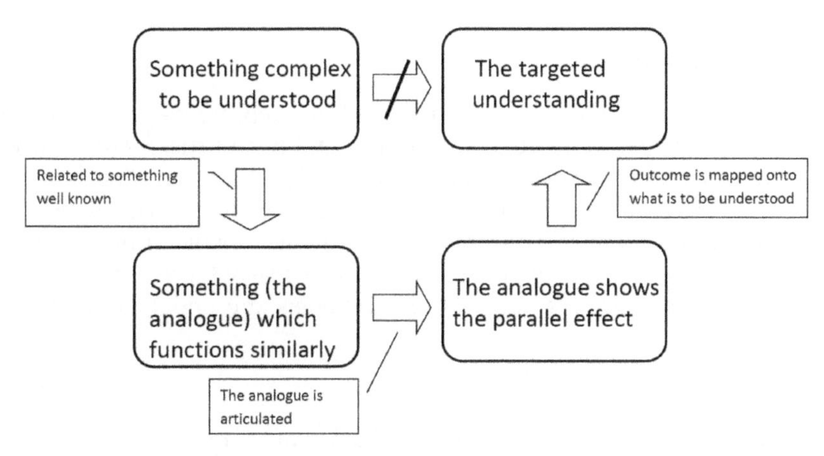

Figure 5.7 Using an analogy to enhance understanding (Newton, 2012).

perception. The teacher presenting the analogy may perceive things differently to the students using the analogy. It is important, therefore, to be aware of the limitations of a chosen metaphor and to use it as an opportunity to discuss the extent to which the metaphor represents the chemical phenomena appropriately. Metaphors are useful from a constructivist viewpoint (Ritchie, 1999) and can become a mechanism to promote discussion and for students to explore their zone of proximal development through social constructivism. In Activity 5.6, we suggest a framework for exploring metaphors with students.

Activity 5.6 – Analogies in Chemistry

This activity aims to develop creative thinking in students by encouraging the development of interesting and unique comparisons for concepts in chemistry.

i) Introduce the ideas of analogies and metaphors with some everyday examples that students may be familiar with such as "the world is a stage".
ii) Present the students with a concept in chemistry such as collision theory and rates of reaction and ask them to come up with their own analogy (*e.g.* rate of reaction is like a football match).
iii) Students explain their analogy to the rest of the class and strengths and limitations are discussed.

A similar task involves looking at everyday objects and exploring how they can be a metaphor in chemistry. For example, in what ways is the pattern of brickwork similar to a cross-linked polymer or an onion similar to an atom. By exploring these ideas, students can develop their mental models and understandings.

5.7 Conclusion

In this chapter, we have explored how multiple representations have been developed to convey knowledge and understandings of the chemical world. These representations highlight alternative perspectives and previously undiscovered inter-elemental relationships. We have demonstrated how the development of these representations has come about through creative thinking. In addition, creative teaching can make use of these representations and analogies to develop students' mental models and understandings.

References

Ainsworth S., (2006), DeFT: a conceptual framework for considering learning with multiple representations, *Learn. Instr.*, **16**, 183–198.

Ainsworth S., (2008), The educational value of multiple representations when learning complex scientific concepts, in Gilbert J. K., Reiner ve M. and Nakhleh M. (ed.) *Visualization: Theory and Practice in Science Education*, Springer, pp. 191–208.

Alexanderdesign, (2019), http://www.allperiodictables.com/AAEpages/aaeDeskTopper.html.

Allred Z. D. R. and Bretz S. L., (2019), University chemistry students' interpretations of multiple representations of the helium atom, *Chem. Educ. Res. Pract.*, **20**, 358–368.

Ashe C. A. and Yaron D. J., (2013), Designing analogy-based simulations to teach abstractions, *Pedagogic Roles of Animations and Simulations in Chemistry Courses*, 367–388.

Beall H., (1999), The ubiquitous metaphors of chemistry teaching, *J. Chem. Educ.*, **76**(3), 366.

Benfey T., (2009), The Biography of a Periodic Spiral, *Bull. Hist. Chem.*, **34**(2), 141.

Chemical Galaxy, (2019), The Chemical Galaxy II, Available at: https://www.chemicalgalaxy.co.uk/index.html.

Coleman S. L. and Gotch A. J., (1998), Spatial perception skills of chemistry students, *J. Chem. Educ.*, **75**(2), 206.

Corradi D. M. J., Elen J., Schraepen B. and Clarebout G., (2014), Understanding possibilities and limitations of abstract chemical representations for achieving conceptual understanding, *Int. J. Sci. Educ.*, **36**(5), 715–734.

Debes J. L., (1969), The loom of visual literacy–An overview, *Audiovisual Instr.*, **14**(8), 25–27.

DePeip, (2019), https://commons.wikimedia.org/wiki/File:Elementspiral_(polyatomic).svg.

Duit R., Roth W. M., Komorek M. and Wilbers J., (2001), Fostering conceptual change by analogies—between Scylla and Charybdis, *Learn. Instr.*, **11**(4–5), 283–303.

EuChems, (2019), The Periodic Table of Scarcity, Available at: https://www.euchems.eu/euchems-periodic-table/.

Gentner D., (1998), Analogy, in Bechtel W., Graham G. and Balota D. A. (ed.), *A companion to cognitive science*, Oxford: Blackwell.

Gilbert J. K. and Justi R., (2016), *Modelling-based teaching in science education*, Cham, Switzerland: Springer International Publishing, **vol. 9**.

Han M. and Kim H. B., (2018), Elementary Students' Modeling Using Analogy Models to Reveal the Hidden Mechanism of the Human Respiratory System, *Int. J. Sci. Math. Educ.*, 1–20.

Harshman J., Bretz S. L. and Yezierski E., (2013), Seeing chemistry through the eyes of the blind: A case study examining multiple gas law representations, *J. Chem. Educ.*, **90**(6), 710–716.

Heywood D. and Parker J., (1997), Confronting the analogy: primary teachers exploring the usefulness of analogies in the teaching and learning of electricity, *Int. J. Sci. Educ.*, **19**(8), 869–885.

Janet C., (1928), La classification hélicoïdale des éléments chimiques, *Imprimerie Départementale de l'Oise*, Beauvais.

Jensen W. B., (ed.), (2002), *Mendeleev on the Periodic Law: Selected Writings, 1869–1905*, University of Cincinnati.

Kozma R., (2003), The material features of multiple representations and their cognitive and social affordances for science understanding, *Learn. Instr.*, **13**(2), 205–226.

Kozma R. B. and Russell J., (1997), Multimedia and understanding: Expert and novice responses to different representations of chemical phenomena, *J. Res. Sci. Teaching*, **34**(9), 949–968.

Leach M. R., (2019a), Bassett's vertical arrangement, Available at: https://www.meta-synthesis.com/webbook/35_pt/pt_database.php?PT_id=62.

Leach M. R., (2019b), Werner's arrangement, Available at: https://www.meta-synthesis.com/webbook/35_pt/pt_database.php?PT_id=64.

Leach M. R., (2019c), The internet database of Periodic Tables, Available at: https://www.meta-synthesis.com/webbook/35_pt/pt_database.php.

Lucas B. and Spencer E., (2017), *Teaching Creative Thinking: Developing learners Who Generate Ideas and Can Think Critically*, Crown House: Publishing Ltd.

Merrill M. D., (2002), Knowledge objects and mental models, *Instr. Use Learn. Objects*, 261–280.

Newton D. P. (2012) *Teaching for Understanding*, London: Routledge.

Nyachwaya J. M. and Gillaspie M., (2016), Features of representations in general chemistry textbooks: a peek through the lens of the cognitive load theory, *Chem. Educ. Res. Pract.*, **17**(1), 58–71.

Oblinger D. G. and Oblinger J. L., (2005), *Educating the Net Generation*, Boulder, CO: Educause.

Oliver-Hoyo M. and Babilonia-Rosa M. A., (2017), Promotion of spatial skills in chemistry and biochemistry education at the college level, *J. Chem. Educ.*, **94**(8), 996–1006.

RCSB, (2019), The Research Collaboratory for Structural Bioinformatics (RCSB) Protein Data Bank (PDB). Available at: http://www.rcsb.org/.

Ritchie S. M., (1999), The craft of intervention: A personal practical theory for a teacher's within-group interactions, *Sci. Educ.*, **83**(2), 213–231.

RSC (Royal Society of Chemistry), (2019), Periodic Table. Available at: https://www.rsc.org/periodic-table.

Seaborg G. T., (1985), The Seaborg/AAE papers, Available at: http://www.allperiodictables.com/AAEpages/SeaborgPapers/SeaborgPapers.html.

Scerri E., (2011), *The Periodic Table: A Very Short Introduction*, Oxford: Oxford University Press.

Seufert T., (2003), Supporting coherence formation in learning from multiple representations, *Learn. Instr.*, **13**, 227–237.

Stewart P. J., (2004), A new Image of the Periodic Table, *Educ. Chem.*, **41**(6), 156–158.

Stewart P. J., (2010), Charles Janet: unrecognized genius of the periodic system, *Found. Chem.*, **12**(1), 5–15.

Sweller J., (1994), Cognitive load theory, learning difficulty, and instructional design, *Learn. Instr.*, **4**(4), 295–312.
Treagust D. F., Harrison A. G. and Venville G. J., (1998), Teaching science effectively with analogies: An approach for preservice and inservice teacher education, *J. Sci. Teach. Educ.*, **9**(2), 85–101.
Turner K., (2019), Reasons to craft your own molecular models, *Educ. Chem.*, **56**(2), 24.
University of Colorado, (2019), https://phet.colorado.edu/.

CHAPTER 6

Storytelling

Storytelling is central to human existence; we listen to many stories every day of our lives. Whether it is through books, TV programmes, movies, newspapers reporting celebrity gossip, political or legal events, sporting events or simply chatting with friends over a cup of coffee. We even construct and reconstruct stories in our dreams (Gottschall, 2012). The development of language in human evolution enabled the telling of stories for sharing of social information (Dunbar *et al.*, 2007) and moral tales of right and wrong in order to maintain cohesion and cooperation within a tribe. Elders told stories about ancestor heroes, quests and magic (Wilson, 2017) that enabled children to make sense of their world. Today, nations are defined by their stories of great victories or defeats and shared values and characteristics.

The mind creates the world around us and is a story processor (Haidt, 2012). Stories are powerful tools for organising, storing, describing and communicating knowledge, and are usefully thought to stimulate supposedly unrelated areas of the brain (Sabatinelli *et al.*, 2006) – an essential part of creative thinking. Any storyteller is faced with the challenge of grabbing and holding the attention of the audience. A well-constructed story has a greater chance of achieving this and conversely, a poorly constructed story will quickly lose the interest of its audience. The creative chemistry teacher recognises the value of developing the skill of storytelling and incorporates the principles within the curriculum and in lesson design. Ball (2001) demonstrated how to craft stories about molecules as engaging as any human story. In this chapter, we will consider the important characteristics of effective storytelling and their incorporation within chemistry teaching to support creative thinking in the teacher and the student. Stories can set the chemistry in a new context which helps students make new mental connections.

Advances in Chemistry Education Series No. 4
Creative Chemists: Strategies for Teaching and Learning
By Simon Rees and Douglas Newton

Published by the Royal Society of Chemistry, www.rsc.org

6.1 Storytelling in Science Teaching

Bruner (1991) made a theoretical justification for using stories in science classes. He pointed out that science discourse naturally tends to rely on a paradigmatic, logical mode rather than a narrative temporal mode, more familiar in a story. Despite the advantages of using stories and narratives and the ubiquitous nature of stories, academic disciplines tend to prefer logical paradigmatic reasoning because it is considered more "scientific" (Jonassen and Hernandez-Serrano, 2002). Peleg *et al.*, (2017) argue that the use of stories can increase students' situational interest (a state of positive emotion and heightened concentration (Hidi and Renninger, 2006)) particularly in the context of inquiry-based learning.

There is limited research on the impact of storytelling in science teaching. Some studies indicate that storytelling activities may improve understanding of, for example, electricity (Braund, 1999) and mixtures and solutions (Arieli, 2007) but not factual recall (Metcalfe *et al.*, 1984; Ødegaard, 2003). Erduran and Pabuccu (2015) also demonstrated the value of stories for developing argumentation.

Science teachers' education background means that it is likely they have limited experience or understanding of narrative based pedagogies and may lack confidence in implementing such strategies in the classroom (Alrutz, 2004). Some science teachers, however, even without formal training, spontaneously use storytelling activities (Dorion, 2009). Peleg *et al.*, (2017) identified authenticity and truth as an issue when science teachers used fictitious stories to frame a scientific mystery. These teachers felt that students would always be expecting them to provide facts and expound accepted beliefs. Some teachers reported that stories ignited interest and they felt that using stories would engage a wider group of students and increase motivation.

6.2 Unexpected Change

The mind is attuned to detecting change. Senses collect information from the environment such as temperature or light levels and send signals to the brain. Consistent and constant stimuli come to be ignored. A mouse hiding in the corner of a room, for example, may remain undetected until it suddenly moves and the change is detected. The mind seeks to manage, control and understand its environment. Unexpected change represents a loss of control and possible threat. It may also be a harbinger of hope or promise. Change promotes curiosity; 'what does this change mean? Is it good or bad?'

Storytellers create moments of change that catch the attention of the protagonists and, by extension, the audience. Aristotle argued that a turning point is one of the most powerful moments in drama. Can you recall the last time you shared with someone how much you enjoyed a TV programme with twists and turns and that unexpected twist at the end? Stories often incorporate change in the very first sentences. For example, "That Spot! He hasn't eaten his supper. Where can he be?" (Eric Hill, 'Where's Spot?'

(Hill, 2009)). In other instances, the hint of change to come is sufficient to capture the reader's interest. For example, Harry Potter and the Philosopher's Stone opens with "Mr and Mrs Dursley, of number four Privet Drive, were proud to say that they were perfectly normal, thank you very much" (Rowling, 2015). The sentence hints that this normality may be under threat.

The threat or possibility of change is an effective way of arousing curiosity. As Alfred Hitchcock said "there is no terror in the bang, only in the anticipation of it". The audience is engaged and their dread heightened as they are unsure what is coming. Building this sense of anticipation is not restricted to fiction. Karl Marx, in his communist manifesto, begins "A spectre is haunting Europe – the spectre of communism" (Marx and Engels, 2002).

Chemistry teaching rarely provides storylines as threatening as one of Hitchcock's films but it is possible to exploit this principle – sowing the seeds of curiosity in the students' minds. For example, consider an experiment to determine the empirical formula of magnesium oxide. This investigation commences with a demonstration of burning magnesium, producing a grey ash product. If the students are asked whether the mass of the product is less or more than the original magnesium, many will say that it is less based on the appearance of the ash. However, when the magnesium is carefully burnt in a crucible and reweighed the *increase* in mass is an unexpected change. This is chemistry teaching applying the storyteller's strategy to promote curiosity and engage the students in discovering the explanation.

Michael Faraday used the same strategy in his lectures. For instance, when holding a lit taper to the "smoke" a couple of inches from the wick of an extinguished candle, the audience observes the candle relighting. This is unexpected because the flame was not touching the wick, so how can the "smoke" catch fire? This unexpected change prompts curiosity and the search for an explanation. This is an effective way of challenging misconceptions in the students' current understanding. Can you think of examples of unexpected change in your own teaching or new ways you could incorporate this principle? Examples include states of matter when cornflour in water can behave both like a liquid and a solid or carbon dioxide – perceived to be a gas but then demonstrated as a solid, dry ice. This is then added to water which appears to boil but is freezing cold! Or establishing the physical properties of ionic and covalent compounds but then discovering exceptions such as sucrose dissolving in water. Or discovering that water is not simply H_2O, but exists as a combination of ions in equilibrium, and the pH of pure water is not 7 at different temperatures although it is still neutral. These examples illustrate how students can experience an unexpected change in their perception and understanding of the world around them, leading to cognitive dissonance and promoting curiosity for an explanation.

6.3 Curiosity

Incorporating unexpected change may be difficult to achieve at times but it is not the only way to arouse students' curiosity. The human mind seeks to understand and control. Between the ages of two and five, we ask around 40 000 explanatory questions about the world around us (Leslie, 2014). What is chemistry if not the human endeavour to develop better understandings of the world around us? The skill of the storyteller and the creative chemistry teacher is to excite these inquisitive instincts without telling the audience everything about them. Michael Faraday tapped into these instincts in his public lectures, asking questions, encouraging new perspectives and revealing understanding but always with a sense of what will be revealed next.

Human minds become curious when presented with incomplete information and the desire to fill the gaps (Golman and Loewenstein, 2018). As more of a story is revealed, the more we want to know and resolve the mystery. Kidd and Hayden (2015) describe curiosity as shaped like a lowercase 'n'. It is at its weakest (the bottom left of the "n") when people have no idea about the answer to a question and they are convinced of this fact. This statement should resonate with every chemistry teacher and highlights the importance of asking questions that are not well beyond the students' actual or perceived understanding. Equally, curiosity is dampened if the subject is complex as when students are required to operate between the macroscopic, sub-microscopic and symbolic levels (Johnstone, 1991). Curiosity is at its highest at the peak of the "n", where people think they have some idea but are not quite sure. This is apparent in chemistry teaching when the students have begun to develop some understanding and become curious to find out more. Curiosity then reduces again (to the bottom right of the "n") as the story reaches its resolution. In the case of chemistry teaching, this may occur when the students feel they have acquired understanding and there is not more to know. Within the restrictions of a syllabus, for example, students may be secure in their understanding required for an examination and there is no motivation to go further.

Loewenstein (1994) identifies four different ways of involuntarily inducing curiosity. These are:

1. the posing of a question or presentation of a puzzle;
2. exposure to a sequence of events with an anticipated but unknown resolution;
3. the violation of expectations that triggers a search for an explanation;
4. knowledge of possession of information by someone else.

These curiosity attributes are employed in detective fiction such as Sherlock Holmes. The story poses a puzzle, there is a sequence of events with an anticipated but unknown resolution, red herrings create false trails and someone knows "who dunnit" but the reader does not. All four of these attributes are relevant to the work of the creative chemistry teacher and should be cornerstones of curriculum delivery.

In science teaching, the application of these principles is exemplified by projects like Fusion Science Theater (see Case Study 6.1) and the Teaching Enquiry with Mysteries Incorporated (TEMI) project. In these projects, students are presented with scenarios that lead to cognitive dissonance, generating curiosity, situational interest and a story to solve the mystery. For example, the TEMI project (Peleg, Katchevich, Yayon, Mamlok-naaman, Dittmar, & Eilks, 2015) introduces students to hydrophobic sand that behaves in a different way to what the students would normally observe. This then requires the students to engage in the story to solve the mystery.

Case Study 6.1 – Fusion Science Theater

Fusion Science Theater (2019) demonstrates the effective application of story in science teaching through a range of science shows focusing on basic science concepts.

The shows use the following storytelling principles:

i) Characters

The audience relate to and build an emotional connection with the characters in the play, making the concepts more personal and compelling.

ii) Dramatic question

The shows use a dramatic question to promote curiosity such as "will it light?" – an exploration of conductivity in different solutions. The audience is asked to vote on possible answers to the question at the beginning and are then engaged with wrestling with the problem as the performance progresses.

iii) Revelations

Over the course of the play, information is revealed that the audience must piece together to answer the question. This requires the active involvement of the audience in asking questions and participating in role play activities, such as modelling the movement of ions in a solution, to help visualise the sub-microscopic world of ions and molecules.

iv) Resolution

The audience are asked to vote once more about what they now think is the right explanation for the problem. The answer to the question is resolved at the end with a demonstration that elicits an excited response, not due to spectacle, but because the explanation has now been revealed and the story comes to a close.

Arousing curiosity is easily achieved when something is a matter of life or death but, as Loewenstein (1994) says, this can also apply to questions of seemingly little significance. Responding to "clickbait" on the internet taps into people's curiosity. Even though five minutes ago someone never knew that they were actually curious about the top ten celebrities that look like their dogs! Stories such as the "ketchup conundrum" (Gladwell, 2004), where a detective seeks to solve the mystery of why it is so hard to make a tomato sauce to rival Heinz, are crafted to maintain the reader's interest in what could otherwise be considered a trivial and uninteresting subject. This is the sign of the master storyteller, where, through skilful use of storytelling techniques, the audience are engaged in a subject that would otherwise have passed them by.

Information gaps are central to the narrative of science fiction programmes, such as "Stranger Things". The writers create a mysterious world, "the upside down", which is slowly revealed as the story progresses. In the same way, creative chemistry teachers are authors of their students' chemical imaginations, which develop as the course progresses. J. J. Abrams said "mystery is the catalyst for imagination" and creative chemistry teaching can employ this principle to spark students' imaginations.

One challenge when incorporating these principles in lessons is the received wisdom of presenting clearly defined learning outcomes at the beginning of a lesson. If these learning outcomes are too explicit and reveal too much of the story at the beginning then curiosity is lost. Imagine reading a book that begins by telling the ending! When promoting films, the stars have to promote curiosity carefully without revealing too much and quenching someone's desire to go to the cinema. The same should apply to learning outcomes presented at the start of a chemistry lesson. They should be designed to promote curiosity with the potential for change in how the students will perceive their physical world.

6.4 Constructing Models of the World

The world we experience with all its sights, sounds, smells and tastes is constructed in our minds as a mental activity. The chemical world of atoms, molecules and reactions is, similarly, a construct based on individual experience. When a student walks into the laboratory, their mind predicts what the scene should look and feel like, and generates a mental scene based on its prediction (Cook, 2015). This scene is then shaped by electrical signals received by the senses and a perceived reality is constructed. The mind uses the pulses to create a colourful set in which to play out our lives and the models of the world we construct.

When words on a page describe gnarled old trees in a forest, the mind will see the trees in the forest and model the scene. The more skilful the writing, the more vividly we see this imagined world (see, for example, the description of the painting in Section 6.8). In the same way, creative chemistry teachers are authors of their students' imagined worlds. Chemistry educators tell the stories to enable students to visualise this sub-microscopic world beyond their immediate experience.

6.5 Detail

An important part of mental model making is the use of specific detail. Precise and specific descriptions lead to precise and specific models.

Summerfield *et al.*, (2010) state that three specific characteristics of an object should be described such as: the dark blue carpet or the orange striped pencil. Applying this principle to chemistry teaching, we can think of similar examples, such as: chlorine is a pale, green gas, or sodium is a soft, grey metal. This is easily achieved at the macroscopic level but is more difficult at the sub-microscopic level. For example, the phrase "an atom consists of electrons, protons and neutrons" offers no descriptive value and does not help the mind construct a model of this reality. If we do not know what electrons, protons and neutrons are like then how can we imagine what an atom is like? At the sub-microscopic level, we have to be creative in our approaches to enable students to construct their own imagined worlds. It is important to choose consistent language with appropriate precision and detail.

Close your eyes and imagine that you have shrunk so small that you are sat on the surface of a proton in the nucleus of an atom. Look up to the "sky" and describe what you can see. Each of us will have our own unique world that is not easy to visualise and is even harder to describe. A world in our mind that we believe exists but no one has seen. Those with a strong background in chemistry or physics will construct a complex world based on years of accumulated knowledge and understanding. For the novice, however, this represents one of the greatest opportunities in engaging with the subject. Do we spend enough time creating a strong and exciting narrative for students' imaginations? Figurative representations such as dots and crosses on a page, or unfamiliar notations such as $1s^2$ do little to enable a student to construct a tangible and believable atomic mental model. These representations were developed by expert chemists with highly developed mental models to enable them to think about the atomic world. For the novice chemist, however, they are unfamiliar and tell them little. The creative chemistry teacher uses the storytelling principles of curiosity and precise and specific detail to help the students to be curious about and begin to imagine this new world. There should be more emphasis placed on descriptive narratives so that students can form tangible mental sub-microscopic models.

Popular TV programmes about astronomy take the audience on a journey of imagination to the most distant reaches of the Universe. The audience is required to imagine a universe of infinite size and alien worlds light years away that are very different from our own. In order to do this, the storyteller will use evocative and descriptive language often taken from some of the Earth's most extreme and hostile environments that offer some similarities to these distant planets. This helps the audience to create mental models in their imagination. Often, the accuracy of this imagined world in a distant galaxy may be less important than the fact that they have been engaged in the story. In the same way, we can go on a journey in the opposite direction into the sub-microscopic level. Animations and videos (*e.g.* Scaleofuniverse, 2019; Secret Worlds, 2019) can help students build mental models, although these representations of atoms are themselves often limited. The challenge for the creative teacher is to use precise and specific detail to engage students in developing more sophisticated and accurate mental models. Activity

6.1 aims to reflect on the language we use to describe the atom and how we can help students visualise it.

Activity 6.1 – Visualising the Atomic World

This activity aims to show how the imagined world of the atom develops as a student progresses through their education. Careful consideration should be given to the words we use to describe the atom.

Reflect on the model of an atom a student will have at different stages of their education such as up to age 11, 16, 18 and 21 years. Even though you probably do not have experience of teaching students at all these ages it is very useful to think about the story of the atom and how that develops over time.

i) How would you describe the atom at each of these stages? List typical words and phrases used.
ii) How useful are these words to describe this world? Do they enable the student to build a suitable mental model? For example, if you say the size of an electron is 1/1840 that of a proton then you need to have some sense of how small a proton is in order to visualise it. If you had never seen a mouse or an elephant then it is no use saying a mouse is 1840 times smaller than an elephant unless you have some idea of how big an elephant is!
iii) For each of the stages, now construct a description of the atom with appropriate precision and detail to enable someone to build a mental model. Try to develop the most concise, precise and specific description that you can and aim to use this consistently.
iv) A description based on words alone may not be sufficient. What other means of communication such as images, physical models or role play may be helpful? However, be mindful of providing too much information, or using images that could confuse.
v) Consider adding sensory texture to this world. What would it feel like to be at the centre of an atom? What would it sound and smell like? This is where students can be creative in constructing their own mental model. There is no right or wrong answer – no one knows what it would feel like to stand on the surface of a proton. This, however, is the exciting thing, feeding curiosity and imagination and can help the sub-microscopic world become meaningful.

6.6 Character

Storr (2019) argues that it is the central characters of a story who are the most important element that engages an audience. We are interested in other people and their plight as they struggle through life. Characters that are true to life and full of narrative surprise engage the mind as it seeks to understand the world. This links with Mahaffy's (2006) idea of the chemistry education tetrahedron which emphasises the human element of chemistry education. As Newton (1988) pointed out, there are three Ps in science: products, processes and people. Humanising science by putting the people back in adds to the perceived relevance, and to student engagement.

However, a scan through any chemistry textbook will tend to reveal limited attempts to tell the stories of the people who shaped our chemical understanding. Mendeleev, for example, appears in the ubiquitous black and white photo as a bearded old man. Recognition and explanation are given to his contribution to developing the modern Periodic Table but there is no narrative about the journey that he took to get there. An alternative character is Mendeleev's mother, who struggled to bring up ten children in Siberia after the death her husband. She recognised Dimitri's talent and, determined to provide him with an education, rode with the young boy on horseback for 1200 miles across the Ural Mountains (Crash Course, 2019). Now, the hero of the story is a woman whose foresight, resilience and fortitude enabled the development of the Periodic Table. This adds a human dimension to the story that many more students will find relatable and engaging.

Consider also Michael Faraday, most often depicted as an older man in full Victorian outfit. However, what about the 14 year old boy, brought up in poor surroundings due to his father often being ill, walking the streets of London looking for work. The young Faraday entered a bookbinder's premises and began an apprenticeship that transformed his life. In this way, the remote genius becomes someone students can relate to and a potential role model.

While these stories may not contribute directly to students' understanding of chemical phenomena, they provide the human element of chemistry with characters that come alive and engage. Motivating students to see the relevance of the chemist's life and even leading to some wanting to be a chemist.

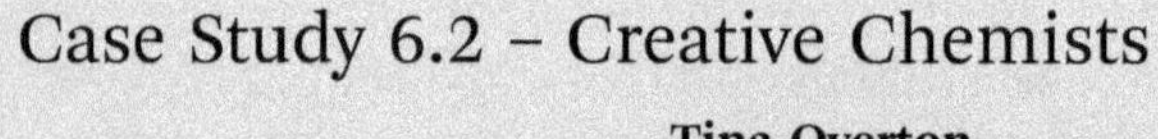

Case Study 6.2 – Creative Chemists

Tina Overton

(Image courtesy of Tina Overton)

Professor Tina Overton is Director of the Leeds Institute for Teaching Excellence. The institute promotes innovation and best practice in student education at the University of Leeds, UK. Tina has taught chemistry to undergraduate students for nearly 30 years. She has taught inorganic, environmental, analytical chemistry as well as a range of transferable skills and career skills modules, both in the UK and in Australia. She has taught on-campus and distance taught students, face to face, online and developed and delivered a mooc. She supports and mentors colleagues to develop their own practice and provides leadership in curriculum and pedagogic development.

How has creative thinking been important in your career?

I have always tried to innovate in my teaching. I am never satisfied with doing things the same way year on year. So I started to change things and innovate very early in my teaching career and creativity played a big part in that. I wanted to deliver different ways for students to learn chemistry and there was nothing readily available. I really wanted to provide students with activities that develop their critical thinking skills and so co-authored a book of problems designed to develop critical thinking. There were no other problems out there, so we had to think very creatively to write several hundred of them, which required thinking about learning chemistry in a very different way. I also became very interested in problem-based learning, of which there were hardly any examples in chemistry in those days. I started to develop them for students, using real life contexts to create problem scenarios to drive the learning of chemistry. Again, developing new and novel resources from scratch required lots of creativity and a different way of looking at how we get students to interact with and learn chemistry. More recently, I authored a mooc, with a colleague at Monash University, on 'How to Survive on Mars'. This involved getting to grips with knowledge outside my comfort zone and designing an engaging experience for our non-science learners and creating a range of interactive resources. Without some level of creativity, the mooc would not have been as engaging or successful.

6.7 Story Stimuli

One way to engage students through storytelling is to incorporate other people's stories as well as our own. We are all exposed to a variety of different stories every day through a multitude of media such as books, TV programmes, social media, songs, paintings, photographs or stories from our own lives. These can be used as the basis for stimulating stories in chemistry as illustrated in the following examples.

6.8 An Experiment on a Bird in an Air Pump

Hanging in the National Gallery in London is an oil painting by Joseph Wright from 1768 (the date is significant) entitled "An Experiment on a Bird in an Air Pump". The painting is available to view online (National Gallery, 2019) with the ability to zoom in on fine details in the painting and discover more of the stories within the painting. The National Gallery provides an evocative description of the painting that is a wonderful example of how descriptive prose can generate a mental model in our mind. If you are not familiar with the painting, then read this description first and compare your mental model with the actual painting. You may even try producing a sketch and comparing that with the original painting.

"It is night in a grand private house: through the window the moon gleams behind a cloud. A travelling lecturer, with all the drama of a magician, fixes his gaze upon us. An audience of men, women and children are gathered around him to watch the experiment he is conducting. The room is lit by a single candle that burns out of sight on the polished table behind a large rounded glass containing a diseased human skull.

The candlelight is diffused through the murky liquid and illuminates the faces of the observers, casting deep, dramatic shadows – every furrow of the lecturer's brow and curl of his silver hair is heightened. A rare white cockatoo has been taken from its cage and placed in a glass container from which the air is being pumped to create a vacuum. The cockatoo convulses in distress as it struggles to breathe. With one hand raised, the lecturer has the god-like power over life or death – he can either expel the air completely and kill the bird or allow the air back in and revive it.

A gentleman times the experiment on his pocket watch, while the youth seated beside him leans in for a closer view. The lecturer points to the ticking watch and, with his other hand on the air valve, looks to us as though the decision is ours. The little girl observes the bird with fascination, but her elder sister cannot bear to watch and covers her eyes. Her father places his arm around her shoulder, perhaps to reassure her, or to explain that all living creatures must one day die. The boy beside the window waits, rope in hand, to see whether he will need to lower the cage for the reprieved bird. The young couple beside the lecturer only have eyes for each other. The elderly man on the right contemplates the skull, lost in thought. The skull and candle are emblems of mortality, but Joseph Wright's painting leaves us uncertain of the outcome for the bird." (National Gallery, 2019)

This description illustrates with words the elements of storytelling that are depicted within the painting; engaging the observer or the reader with a variety of characters to build an emotional connection with. Precise and specific detail is used to promote curiosity and there is the threat of unexpected change.

When using this painting with chemistry classes, the painting is presented with little explanation and the students observe and interpret the scene. In so doing they develop observational and questioning skills, as well as their own stories from the scene. Some students are drawn to the distressed face of the young girl while others may notice detail such as the contents of the jar on the table. Or, looking even more closely, the broken line of the stirring rod in the jar, or the image of the candle depicted around the edge of the jar.

The discussion then develops its scientific focus, considering the nature of a vacuum and why the bird is floundering. Our understanding of oxygen can then be compared with the understanding at this time which is approximately 10 years prior to the discovery of oxygen. This enables stories of scientific discovery to be explored, the understanding of the people in the picture and the development of knowledge.

Furthermore, the exploration of the painting leads to the students recreating their own experiment. Not using real birds, of course, but candles placed in a jar over a tub of water. The students design an experiment to investigate the effect of changing a variable such as the number of candles or size of jar on how long the candle will burn or the height the water will rise inside the jar. The observation of the water rising in the jar enables the use of the storytelling device of unexpected change. Many students may explain the rising water level as a result of the oxygen being consumed inside the jar. However, exploring the chemical equation indicates that carbon dioxide is produced as the oxygen is being used up. Careful observation reveals that the water level only tends to rise once the flame is extinguished. Therefore a different explanation is required.

This storytelling device is the same as that employed by Fusion Science Theater where an initial explanation is shown to be false and an alternative explanation is developed. The students demonstrate creative thinking to apply an alternative concept. The effect of the increased temperature caused by the burning candle may lead to a discussion of pressure changes and an explanation as to why the water rises when the flame is extinguished and the air cools down. This can then be linked to the vacuum pump in the painting and how that reduces pressure. Finally, the experiment with the candle can be related to the bird with a discussion about the two central chemical reactions depicted in the painting, respiration within the bird being synonymous with the combustion reaction in the candle.

From the starting point of the painting, the creative chemistry teacher is able to weave a story with multiple narratives that are rich in both human interest and scientific content. Furthermore, the use of such stimulus materials has the potential to bring in other disciplinary areas, such as art and history, and engage a broader diversity of students in chemistry.

6.9 Fictional Stories

Popular fictional stories in books, movies or TV programmes provide many opportunities for creative chemistry teachers to engage their students. When Breaking Bad (a TV series depicting the exploits of a chemistry teacher turned illegal drug dealer) was popular, chemistry teachers were regularly engaged by their students in conversations about the programme and the science involved. The creative chemistry teacher makes the most of these opportunities, finding parts of the story and linking them to relevant parts of the curriculum. This engages the students because they are emotionally invested in the story, and the chemistry lesson will enable them to deepen their understanding. For example, if the latest Spiderman movie has been released, the creative chemistry teacher can ask questions such as: is it possible to produce a spider's thread that is strong enough for a person to swing on (link with polymers) or climb vertical walls (link with intermolecular forces)? The Secret Science of Superheroes (Lorch and Miah, 2017) explores the science of superheroes with a series of articles that can be used as a story stimulus.

This is only applicable as long as there is a current interest in the story. Today, no students mention Breaking Bad. It is no longer relevant and something else has taken its place. The potential for engagement is much reduced, and it is important to find stories that are current. The transient appeal of these stories is also an important reason not to invest significant amounts of time and resource into designing teaching activities around a popular theme. A curriculum heavily designed around, rather than judiciously enriched by, a popular TV programme is at risk of becoming quickly obsolete. Activity 6.2 is a suggested process to explore the potential links with popular culture.

Activity 6.2 – Finding Links with Popular Narratives

The aim of this activity is for you to identify chemistry curriculum links with popular culture relevant to your students. It is ideally suited to a group of creative chemistry teachers working together.

i) Find out what cultural phenomena are currently popular with your students. This could be in books, TV programmes, movies, music or social media. These may provide contentious characters which are popular with some students but not others.
ii) As a group, engage with the content and identify as many possible links to the chemistry curriculum as you can. Think divergently and suspend criticism at this stage.
iii) Now, thinking convergently, identify those links that offer the most potential for your teaching. Do they provide a strong link with the curriculum or is it tenuous and forced? Do they address common misconceptions or the opportunity to explore challenging areas of the curriculum?
iv) This process should be repeated each year and outdated examples discarded. In following this process you will identify novel and purposeful teaching activities – the principle outcome of creative thinking.

Fiction can offer many opportunities to link with the chemistry curriculum and promote literacy. Stories such as George's Marvellous Medicine (Dahl, 2016), Itch (Mayo, 2012) or Harry Potter (Rowling, 2015) offer opportunities to engage students in reading as well as chemistry. The creative chemistry teacher will team up with literacy leads within their schools to promote relevant books and devise lessons that help bring the stories to life *and* develop scientific understanding.

6.10 Our Own Stories

An alternative is to engage students in stories carefully selected from our own lives. Biographies and films based on true stories are immensely popular. They enable the audience to relate better to the person as they build an emotional connection and develop a better understanding. The creative chemistry teacher may use this narrative to incorporate aspects of their own lives, whether it is their own interests or dramatic events (see Chapter 7, Example 7.1). These experiences can be used as starting points to discuss the chemistry that is linked to that experience. Peleg *et al.* (2017) reported that teachers felt more comfortable telling a story as if it really happened to them to make the story more credible. There may also be local stories about other people that provide relatable characters for your students that can be linked with the curriculum. For example, at Durham University, there is the story of the chemistry student and their love of baking. She set up her own cupcake company and then went on to obtain a significant role as a food scientist with a large multinational company.

6.11 Conclusion

This chapter has explored the components of effective storytelling and how to incorporate these ideas into chemistry teaching. It aimed to raise awareness of the skills of a storyteller and rethinking approaches to engaging students. This includes incorporating unexpected change, nurturing and sustaining curiosity and finding and developing engaging characters. It is important to remember, however, that these are stories with a pedagogical purpose and are not simply for entertainment without learning. This is something that many chemistry educators, knowingly or unknowingly, have incorporated into their teaching and there is undoubtedly benefits in being more knowledgeable about story construction to develop courses, lectures and individual activities that incorporate these principles.

References

Alrutz M., (2004), Granting science a dramatic license: Exploring a 4th grade science classroom and the possibilities for integrating drama, *Teach. Artist J.*, 2(1), 31–39.

Arieli B. B., (2007), *The Integration of Creative Drama into Science Teaching* (Doctoral dissertation, Kansas State University).

Ball P., (2001), *Stories of the Invisible: A Guided Tour of Molecules*, Oxford: OUP.

Braund M., (1999), Electric drama to improve understanding in science, *School Sci. Rev.*, **81**, 35–42.

Bruner J., (1991), The narrative construction of reality, *Critical Inquiry*, **18**(1), 1–21.

Cook G., (2015), Exploring the mysteries of the brain, *Scientific American*, Available at: https://www.scientificamerican.com/article/exploring-the-mysteries-of-the-brain/.

Crash Course, (2019), The Periodic Table, Available at: https://www.youtube.com/watch?v=0RRVV4Diomg.

Dahl R., (2016), *George's Marvellous Medicine*, Puffin Books.

Dorion K. R., (2009), Science through drama: A multiple case exploration of the characteristics of drama activities used in secondary science lessons, *Int. J. Sci. Educ.*, **31**(16), 2247–2270.

Dunbar R., Barrett L. and Lycett J., (2007), *Evolutionary Psychology*, Oneworld.

Erduran S. and Pabuccu A., (2015), Promoting argumentation in the context of chemistry stories, in *Relevant Chemistry Education*, Brill Sense, pp. 143–161.

Fusion Science Theater, (2019), Available at: http://www.fusionsciencelearning.org/.

Gladwell M., (2004), The ketchup conundrum, *New Yorker*, 6.

Golman R. and Loewenstein G., (2018), Information gaps: A theory of preferences regarding the presence and absence of information, *Decision*, 5(3), 143–164.

Gottschall J., (2012), *The Storytelling Animal: How Stories Make us Human*, Houghton Mifflin Harcourt.

Haidt J., (2012), *The Righteous Mind: Why Good People are Divided by Politics and Religion*, Vintage.

Hidi S. and Renninger K. A., (2006), The four-phase model of interest development, *Educ. Psychol.*, **41**(2), 111–127.

Hill E., (2009), *Where's Spot*, Puffin Books.

Johnstone A. H., (1991), Why is science difficult to learn? Things are seldom what they seem, *J. Comput. Assisted Learn.*, 7(2), 75–83.

Jonassen D. H. and Hernandez-Serrano J., (2002), Case-based reasoning and instructional design: Using stories to support problem solving, *Educ. Technol. Res. Dev.*, **50**(2), 65–77.

Kidd C. and Hayden B. Y., (2015), The psychology and neuroscience of curiosity, *Neuron*, **88**(3), 449–460.

Leslie I., (2014), *Curious: The Desire To Know and Why Your Future Depends On It*, Basic Books.

Loewenstein G., (1994), The psychology of curiosity: A review and reinterpretation, *Psychol. Bull.*, **116**(1), 75.

Lorch M. and Miah A. (ed.), (2017), *Secret Science of Superheroes*, Royal Society of Chemistry.

Mahaffy P., (2006), Moving chemistry education into 3D: A tetrahedral metaphor for understanding chemistry. Union Carbide Award for Chemical Education, *J. Chem. Educ.*, **83**(1), 49.

Marx K. and Engels F., (2002), *The Communist Manifesto*, Penguin.
Mayo S., (2012), *Itch*, Corgi Childrens.
Metcalfe R. J. A., Abbott S., Bray P., Exley J. and Wisnia D., (1984), Teaching science through drama: An empirical investigation, *Res. Sci. Technol. Educ.*, **2**(1), 77–81.
National Gallery, (2019), Available at: https://www.nationalgallery.org.uk/paintings/joseph-wright-of-derby-an-experiment-on-a-bird-in-the-air-pump.
Newton D. P., (1988), *Making Science Education Relevant*, Kogan Page.
Ødegaard M., (2003), Dramatic Science. A Critical Review of Drama in Science Education, *Stud. Sci. Educ.*, **39**(1), 75–101.
Peleg R., Katchevich D., Yayon M., Mamlok-naaman R., Dittmar J. and Eilks I., (2015), The magic sand mystery, *Science in School*, 32.
Peleg R., Yayon M., Katchevich D., Mamlok-Naaman R., Fortus D., Eilks I. and Hofstein A., (2017), Teachers' views on implementing storytelling as a way to motivate inquiry learning in high-school chemistry teaching, *Chem. Educ. Res. Pract.*, **18**(2), 304–309.
Rowling J. K., (2015), *Harry Potter and the Philosopher's Stone*, Bloomsbury Publishing.
Sabatinelli D., Lang P. J., Bradley M. M. and Flaisch T., (2006), The neural basis of narrative imagery: emotion and action, *Prog. Brain Res.*, **156**, 93–103.
Scaleofuniverse, (2019), *The Scale of the Universe 2*, Available at: https://scaleofuniverse.com/.
Secret worlds, (2019), *Secret worlds: the universe within*. Available at: https://micro.magnet.fsu.edu/primer/java/scienceopticsu/powersof10/.
Storr W., (2019), *The Science of Storytelling*, William Collins.
Summerfield J. J., Hassabis D. and Maguire E. A., (2010), Differential engagement of brain regions within a 'core' network during scene construction, *Neuropsychologia*, **48**(5), 1501–1509.
Wilson E. O., (2017), *The Origins of Creativity*, Liveright Publishing.

CHAPTER 7

Performance and Drama

In a way, every teacher is a performer, standing up in front of an audience and engaging them in the subject to hand. In this chapter, we build on the theme of storytelling and focus on creative approaches to performance and drama based pedagogies within chemistry teaching. This includes the elements of performance for the teacher as well as the pedagogical potential of performance and drama based activities to engage students and develop chemical understandings. We only have to reflect on the popularity of extra-curricular drama clubs to recognise that drama based activities can engage a wide diversity of students and those who may struggle to find interest in chemistry. In addition, this has the potential to help students make new and unexpected mental connections.

7.1 Drama in Science

Drama based pedagogies are primarily collaborative and improvisational. They can promote higher level thinking skills such as analysis, synthesis and evaluation (Harvard-Project-Zero, 2019; Wagner and Barnett, 1998) as well as imagination and creative thinking. They provide excellent opportunities for dialogical teaching and knowledge development through negotiation of meaning rather than non-interactive and authoritative discourse. The role of the teacher is to guide and model discourse, promote metacognition and an inclusive learning environment. Meaning and understanding are negotiated between the group and the teacher. Furthermore, there are consistent findings in the literature of high motivation among students, grounded in a sense of student ownership and empowerment (Ødegaard, 2003).

A pedagogical activity becomes drama when the participants are required to "behave as if their world is different from reality" (Anderson, 2004). The participant is transported into a situation that requires imagination and belief. Dorion (2009) describes drama as the enactment of an imagined situation through role play in the human dimension. This imaginary

Advances in Chemistry Education Series No. 4
Creative Chemists: Strategies for Teaching and Learning
By Simon Rees and Douglas Newton

Published by the Royal Society of Chemistry, www.rsc.org

environment must be negotiated within the actor's real physical world. In order to achieve this, participants hold two forms in mind at the same time, or as a state of "double consciousness" (Wilhelm and Edmiston, 1998).

Dorion (2009) recorded a significant variety of drama based activities in science lessons. These may require the participants to assume different human characters such as historical roles (Solomon, 1990) in order to re-create key developments in science or forensic investigations of fictional crimes (Heathcote, 1991). Alternatively, participants may engage in activities to develop understandings of abstract scientific concepts and assume roles such as atoms in a chemical reaction or electrons in a circuit (Dorion, 2009). Some researchers have suggested that drama should be an exploration of the human condition (Somers, 1994) as opposed to role play of non-human concepts. Within science education, researchers have generally regarded role play of scientific concepts as drama (Aubusson and Fogwill, 2006). In interpreting Mahaffy's (2006) human element of chemistry education, the link with exploring the human condition is clear. However, role play activities of chemistry concepts are also engaging the human element and makes the concepts meaningful and memorable. These activities require human engagement both physically and mentally and strengthen the relationship between the student and the chemical phenomenon. Metcalfe *et al.*, (1984) describe these activities as enabling students to develop empathy with non-human entities such as atoms. While we are rightly mindful of anthropomorphising chemical entities, these activities can help students relate to and imagine the chemical nature of the world more deeply.

A significant challenge, of course, is that very few chemistry teachers have any sort of training in drama and may feel uncomfortable participating in such activities. However, it is important to remember that our teaching should be designed to reach out and engage as wide a diversity of students as possible. If a pedagogical strategy has merit then educators have a responsibility to develop their skills and increase their classroom repertoire.

7.2 Performance with a Pedagogical Purpose

Chemistry teaching has a strong tradition for providing opportunities to engage an audience with "whizz bang" science shows. Whether it is these entertaining one offs or regular teaching in the classroom or laboratory, having a good understanding of showmanship and performance is essential. In this section we will explore these elements in the context of effective teaching and learning. That is to say, performance with a pedagogical purpose that goes beyond simply providing memorable "wow" moments of a spectacular demonstration but engages the audience more deeply in order to develop chemical understandings.

7.2.1 Faraday – Chemistry's Greatest Showman

Aside from his scientific achievements and discoveries, Michael Faraday was the greatest science communicator, or showman, of his time. With the establishment of the Royal Institution public lectures in London in the 1820s, he was a founder of the modern public engagement lecture. Were he alive today, he would most likely be a TED talk superstar.

Faraday's lectures did not rely on the spectacular to engage his audience. His most famous series of public lectures were entitled "On the Chemical History of a Candle". Imagine giving a public lecture with that title? Would the audience be queuing around the block to get in? His skill was to recognise the value of ubiquitous and everyday chemical phenomena as the hook to engage his audience and reveal the fascinating stories behind them.

Sadly, or intriguingly, Faraday's performances pre-date video cameras and we are unable to observe his performances today. What would it have been like to sit in the audience of one of his lectures? Imagine being in the audience, crammed into the confined space of the lecture theatre at the Royal Institution – the sense of excited anticipation, the unknown that was about to be revealed.

Fortunately, the complete transcript of his lectures is available (Faraday, 1865) and they provide valuable insights into the art of performance and engaging an audience. For example, early in his first lecture Faraday states:

> *"…so wonderful are the varieties of outlet which it offers into the various departments of philosophy. There is not a law under which any part of this universe is governed which does not come into play and is touched upon in these phenomena."*

He is explaining to the audience the connection between the everyday and possibly mundane phenomenon of a candle burning and the laws that govern the entire universe. The importance of the burning candle is elevated from an everyday phenomenon to something of fundamental importance. He is promoting creative thinking, encouraging the audience to start to consider what they are observing from a new perspective. He also stimulates curiosity and hooks the audience to wanting to know more. Key to an effective narrative, in any context, is to create scenarios and questions that capture the audience's interest and its desire to find out what happens next. What has a candle got to do with the laws of the universe?

In his own "Advice to a lecturer" Faraday (1960) provides valuable guidance that is as pertinent today as 150 years ago. He recommends, in his poetic language:

> *"A flame should be lighted at the commencement and kept alive with unremitting splendour to the end."*

In his lectures on the chemical history of a candle he is lighting both a physical flame and a metaphorical flame of curiosity. Faraday goes on to say:

> *"..And, though I stand here with the knowledge of having the words I utter given to the world, yet that shall not deter me from speaking in the same familiar way to those whom I esteem nearest to me on this occasion."*

In this quote, he is highlighting a golden rule of performance – know your audience. Faraday recognises the importance of the language he used to explain the phenomena; he knows that if he uses words that the audience are not familiar with, and does not explain them, then the performance will fail as the audience becomes confused and loses interest. He also recognises the challenges of engaging a general audience in the philosophy of science, as a way of easing the journey he states:

> "*. . for though to all true philosophers, science and nature will have charms innumerable in every dress, yet I am sorry to say that the generality of mankind cannot accompany us one short hour unless the path is strewed with flowers.*"

In moving on to consider the nature of the candle, he says:

> "*How is it that this solid gets there, it not being a fluid? Or, when it is made a fluid, then how is it that it keeps together? This is a wonderful thing about a candle.*"

Faraday is not simply launching into an explanation of the phenomenon and demonstrating his own understanding but, rather, is encouraging his audience to develop curiosity and think creatively about a phenomenon they may have probably taken for granted in the past. He also demonstrates another golden rule of performance – enthusiasm. By describing a candle as "wonderful" he has elevated its status to an object of wonder, both for himself and his audience. If the audience are engaged with the showman and consider him credible, they will start to believe that there is indeed something wonderful about a seemingly simple candle. Nevertheless, his style was not overly enthusiastic, but very natural. A contemporary commented:

> "*his manner was so natural, that the thought of any art in his lecturing never occurred to anyone*" (Faraday, 1960).

In his advice to a lecturer, he highlights the importance of delivery that is "*slow and deliberate, conveying ideas with ease. . . infusing them with clearness and readiness into the minds of the audience*".

Faraday goes on to reveal something of the nature of science and scientists when he says:

> "*. . ..and I hope you will always remember that whenever a result happens, especially if it be new, you should say, "What is the cause? Why does it occur?" and you will, in the course of time, find out the reason.*"

As much as the audience may have come to learn about the science, they have also come to learn more about those who engage with it. Faraday had achieved something of a celebrity status and audiences were intrigued to find out more about him and his scientific approach to make new discoveries. Through simple experiments, he demonstrates this scientific philosophy. For example, he says:

> "*I will blow out one of these candles in such a way as not to disturb the air around it by the continuing action of my breath; and now, if I hold a lighted*

taper two or three inches from the wick, you will observe a train of fire going through the air till it reaches the candle." And "Suppose I take this candle, and hold a piece of paper close upon the flame, where is the heat of that flame? Do you not see that it is not in the inside? It is in a ring,".

These simple experiments reinforce his earlier statement. A new result has happened. What is the cause? Why does it occur? Faraday is not relying on a sophisticated or spectacular demonstration but rather the simplest of experiments that anyone could do. These lead to some surprising observations that promote curiosity and the search for an explanation.

Using the original lecture transcripts as inspiration, Case Study 7.1 describes a recreation of Faraday's lectures for a 21st century audience. The recreation applies Faraday's ideas and uses multiple narratives to engage the audience and develop chemical understandings.

Case Study 7.1 – Reimagining Faraday

A 14 year old boy walks down a street in London and sees an advertisement for an apprentice in a bookbinder's window. He enters the shop to enquire and is offered the position. As he spends the next seven years binding the volumes, he becomes increasingly interested in the words contained within; igniting his curiosity in the world around him.

So begins a recreation of Faraday's lecture on the chemical history of candle. The audience's interest is not initially sparked by the science but rather by the person – recounting Faraday's early life and how this led to becoming one of the world's greatest scientists. The personal narrative incorporates the human element of chemistry teaching (Mahaffy, 2006) as well as social history. The performance progresses on to recreating Faraday's lecture on the chemical history of a candle, assuming his character, complete with cravat and coat tails (see picture). The audience are invited to imagine what it would have been like to have been present at one of his lectures.

The audience is actively engaged in developing questions and ideas leading to an explanation of the phenomenon. Assumptions are challenged and simple experiments undertaken to promote creative thinking. Using the scientific narrative of combustion, the performance progresses on to a number of other experiments (exploding custard powder, the whoosh bottle and dry ice) to explore the chemical phenomena.

The scientific narrative is brought to a resolution with the development of the scientific explanation. The story, however, has actually only just begun with the audience invited to ask any questions that they have. This leads to a lively question and answer session, with the children asking creative and insightful questions. It also provides the stimulus for further investigations that they can undertake.

This case study is an example of where multiple narratives are weaved together to create an engaging performance to explore chemistry. As one observer commented:

> "*A huge thank you for your Faraday performance today – it was really amazing, and the children absolutely loved it. I sometimes used to say in primary school that we had moments of awe and wonder, and your presentation was one of those moments*".

His ideas of great performance are reflected in contemporary discussions on lecturing and presentation performance. Bailey (2008), for example, argues for the importance of "teacher centred teaching" rather than "student centred learning". He focuses on the qualities of good teaching and, in terms of good performance, he highlighted: speaking clearly, explaining things well, involving the students, being enthusiastic, structuring content to engage and confidence in front of an audience.

Chris Anderson, the developer of TED Talks, uses similar imagery when he talks of lighting a fire of ideas in people's minds and the importance of the idea that you wish to share (Anderson, 2016). From this, he develops the idea of a throughline – the connecting theme that ties each narrative element together. This is evident in plays, films and novels as well as performances. In common with Faraday, he highlights the importance of using a shared language with the audience. He says:

> "*If you start only with your language, your concepts, your assumptions, your values, you will fail. So instead, start with theirs. It's only from that common ground that they can begin to build your idea inside their minds*".

Fundamental to a great performance is telling a good story (see Chapter 6); it is through listening to stories that we build empathy with the characters and find ourselves immersed in their thoughts and emotions. The listeners care about the outcome and their attention is held (Anderson, 2016). He identifies four key elements to a successful story:

- base it on a character your audience can empathise with;
- build tension, whether through curiosity, social intrigue, or actual danger;
- provide the right level of detail for the audience to imagine the scene;
- end with a satisfying resolution, whether funny, moving or revealing.

Activity 7.1 is intended to encourage you to consider about how you can develop a story.

Activity 7.1 – Developing the Story

This activity is designed to develop the elements of performance and narrative within the chemistry curriculum.

i) Think of an area of the curriculum where the students have difficulty.
ii) Using Anderson's four key elements, develop a story that could be incorporated into this curriculum area.
iii) What sort of character would the audience identify with?
iv) How can tension be built into the story? What questions can be asked? What is unknown?
v) How can you help the audience visualise the story without involving too much detail?
vi) How is the story resolved in the end?

This process is demonstrated in the following example.

Example 7.1 – Developing the Story

This example uses the context of a premature baby in an incubator to develop a story that engages students in the curriculum context of acids and bases.

Curriculum area

Acids and bases.

Main character

A premature baby lying in an incubator.

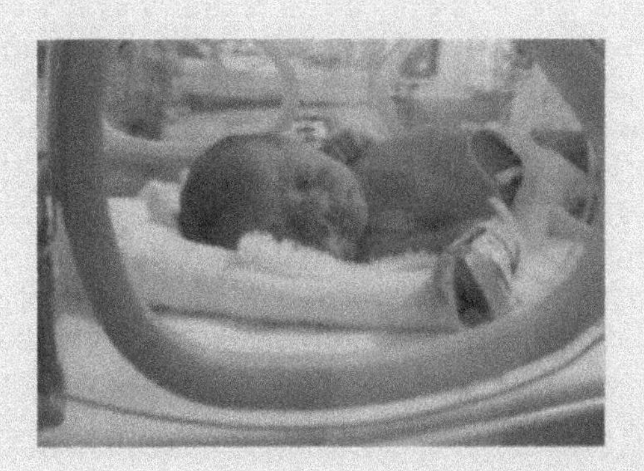

Detail

The fragility of the baby, the scene on the hospital ward, the noise of the machines, the atmosphere and the different people involved such as doctors, nurses and parents.

Build tension

Why is the baby in the incubator? Will the baby survive?
Describe the process of feeding the baby using a syringe attached to a tube entering the nose. Before milk is released down the tube, fluid is sucked up from the tube. This fluid is dabbed on to a small piece of orange paper. What is the paper? Why do they do this?

Resolution

Through discussion with the students, it is determined that the orange paper is universal indicator paper and it is used to test that the tube is going into the stomach, where the acid would turn the paper red, rather than the lungs. If the milk was released into the lungs then the baby could drown.

Notes

Stories such as these are highly emotive. There may be people in the audience whose lives have been affected in some way by the context. The presenter should be aware of this at all times and ensure that the situation is presented and discussed in as respectful and considerate way as possible. Aside from the main curriculum link, there are also many other relevant links to chemistry and wider science curricula within this story.

7.3 Drama and the Human Element of Chemistry

A strong argument for introducing drama activities to the chemistry classroom is their potential for developing empathy in students; the ability to understand the perspectives and emotions of other people. Developing empathy can help students consider moral and ethical issues (Duveen and Solomon, 1994) and also raise self-awareness as students reflect on their own response to different situations (Heathcote, 1991). Chemistry is a subject that some students can find difficult to relate to as the human element is hidden within the abstract world of atoms and reactions and the impersonal nature of the text. Curriculum content also tends to present static knowledge which is then conveyed by the authoritative teacher. Opportunities to engage students in the human element of the subject may feel less relevant and more restricted. However, these activities can achieve broader educational objectives and improve engagement and interest.

Published plays exploring chemistry and the nature of chemists are not common. Carl Djerassi (see Case Study 7.2) was a highly successful chemist who also published several novels and plays exploring a variety of themes within chemistry. He states that:

> *"the fundamental problem with plays focussing on chemistry is that theatre professionals...are threatened, if not actually terrified by the subject. Chemistry, after all, deals with molecules, not people and uses the pictography of chemical formulae and not words"* (Djerassi, 2012).

Case Study 7.2 – Carl Djerassi (1923–2015)

"I feel like I'd like to lead one more life. I'd like to leave a cultural imprint on society rather than just a technological benefit."

Image provided by the Science History Institute (licensed under the CC BY-SA 3.0 licence, https://creativecommons.org/licenses/by-sa/3.0/).

In 1938, Carl Djerassi fled to Bulgaria from Austria at the age of 14 to escape the Nazis and then emigrated with his mother to the United States in 1939. It was here that he studied chemistry and obtained his PhD in organic chemistry from the University of Wisconsin-Madison in 1945. He is credited with leading the team that developed the first oral contraceptive. He had a highly distinguished career as a chemist but also as a novelist and playwright. He wrote several "science in fiction" novels and plays that explore themes such as how scientists work and think as well as scientific ideas and explanations.

Through his life, Djerassi illustrates many capabilities of a creative thinker. In particular his motivation and ability to engage successfully with scientific and artistic spheres. This illustrates an openness to ideas and a willingness to collaborate. Aside from imagining possibilities, he also demonstrated persistence and resilience to achieve his ambitions.

These challenges could equally be applied to chemistry education overall and not just in this particular context. Djerassi adopted the approach of being meticulous with the science in his plays while not overwhelming the audience. One of his most successful plays "Oxygen" was written jointly with Roald Hoffmann to mark the centenary of the Nobel Prize. The play revolves around discussions by the Nobel committee to award a new Nobel Prize – the "retro-Nobel" to honour inventions or discoveries prior to the establishment of the Nobel Prizes in 1901. The action alternates between 1777, a crucial year in the discovery of oxygen, and 2001. The play is historically accurate and explores the fundamental questions relating to discovery in science and the importance to a scientist to be the first to discover something. This play, therefore, is an excellent example of linking historical situations to the present day and exploring themes that remain relevant. The play does not provide a clear resolution as to who eventually was awarded the Retro-Nobel, promoting further discussion.

Other playwrights who have enjoyed success include Jean-Noël Fenwick with his play, "Les Palmes de M. Schutz". Set entirely within a laboratory, the story deals with the discovery of radium by Marie and Pierre Curie. Another example is Stephen Poliakoff's play, "Blinded by the sun", which explores the development of a "sun battery" to split water into hydrogen and oxygen.

However, theatre critics tend not to be excited by the science but more by the human chemistry between the characters. Significant life events for Pierre and Marie Curie such as the sudden death of Pierre and Marie's scandalous affair with Langevin are far more exciting. We may lament that some people struggle to be excited by the chemistry but we should recognise human nature and use it to engage the audience with chemistry where possible.

It is, of course, unrealistic and inappropriate to incorporate entire readings or performances of these plays within the chemistry classroom. However, the creative teacher should be open to finding opportunities to make use of this medium. Perhaps there is the potential for a collaboration between science and drama departments? Could a play be performed and enriched with some related science demonstrations? Excerpts from the plays could be performed within a class to engage the students, or they might devise their own performances (see Activity 7.2).

Case Study 7.3 – Performing Elements

By Chris Thomson, Associate Professor, School of Chemistry, Monash University.

Performing Elements is a performance based learning activity designed for first year university chemistry students. This four week experience has clear links to the organic and inorganic chemistry curriculum, a mix of organic chemistry and inorganic chemistry, articulated through a set of learning outcomes which are quite deliberately a mix of chemistry topics but also several focusing on communication, the social, environmental and ethical responsibilities of chemists and effective teamwork.

Overall, the activity consists of three tasks, over four weeks:

i) Chemical Identities: Each student is allocated a chemical element (*e.g.* iron, gold, nickel) to research before the first session (chemical properties, date of discovery, common uses in society and industry, *etc*). Students then participate in a number of tasks demonstrating this knowledge, such as arranging themselves in chronological order of discovery, ascending atomic number or melting point, or by physically embodying a solid/liquid/gas state depending on a given temperature.

   Students are also assigned a famous chemistry scientist to research and learn about and embody. Vocal coaching tasks were embedded throughout the activity, such as exercises for projecting voice, positive body language, and proper use of a standing microphone. Students then used their chemical personas to introduce themselves to their peers and further develop these important oral communication skills.

ii) Chemical Detectives: This task is designed around the problem solving skills required to determine unknown chemicals from data. Approximately 12 students are assigned the identity of an organic molecule, which is printed and attached to their clothing in full view. Each student is given just a small part of the full data set, and ultimately the task is for the students to work as a team to match the data, compile their evidence, and then determine the identity of the unknown chemical.

   The task is inspired by the well-known game called 'Werewolf'. In this variation, the 12 students and their molecules belong to a molecular 'village' where one of the 12 are secretly assigned as a 'murderer'. The molecular villagers have two minutes to try and identify the structure of the molecule before 'nightfall', after which they must close their eyes. Overnight, the murderer secretly selects one student to be 'killed', and thus removed from the group. Their piece of data leaves with them. The remaining group then continues on the following 'day' to piece the clues together and solve the problem before the next 'nightfall'. The task is ultimately designed not only for students to learn about identifying chemicals from data, but to develop their teamwork skills. This task is typically undertaken during the first and second weeks of the activity.

iii) Chemical Conversations and Storytelling: Groups of three or four students deliver a short performance in front of their peers, which they have written and choreographed themselves. Each group base their performance on either a historical development in chemistry, or a particular piece of chemical knowledge or theory. The key objective is to deliver a performance through which the audience will learn something new about the field. Performances created by students have included dynamic, enthusiastic, but also scientifically accurate themes such as: a mock courtroom scenario debating where certain chemical elements belong in the Periodic Table or the behaviour of electrons inside atoms.

Activity 7.2 – Incorporating a Play into Learning – A Role for Students

i) Decide on an aspect of the curriculum that could provide the opportunity to create a short scene to perform. This might be an historical event related to a key discovery, *e.g.* the scene in Rutherford's laboratory as they undertake the gold foil experiment. Or it could be a scene relating to current events such as a meeting between the climate sceptic president of a country and a group of climate scientists.

ii) As a group, devise a short play to act out this scene.

iii) Ensure that the dialogue includes relevant scientific explanations.

The students should be provided with some guidance on how to construct the scene and the interactions. Example extracts from science plays could be used to illustrate key points such as how scientific explanations are approached.

This activity encourages the students to think creatively to imagine the scene and how the characters would interact.

7.4 Scientific Debate

Public debates have been used for centuries to discuss issues in a way that enables all voices and opinions to be heard and evidence to be considered, hopefully within a respectful and considered environment. There are clearly defined structures to a debate which, if adhered to, enable this process to be followed. The quality of debate that students may observe in their everyday life may not reflect this standard. It is, therefore, all the more imperative to try to teach some effective debating skills. A well organised debate encourages students to develop creative thinking in order to make their case as effectively as possible as well as critical thinking when responding to points made during the debate. In order for this strategy to be effective, it is important to have a good understanding of an effective debate structure, such as:

1. A motion is proposed for the debate such as "this house believes that chemistry is the most important subject to study at school" or "this house believes that carbon is the most important element in the world" or "this house believes hydrogen bonding is the most significant intermolecular force". The motion should make a clear statement that facilitates good opportunities to argue both for and against.
2. Those arguing for the motion are "proposers" and those arguing against are the "opposers".
3. A chair for the debate is appointed and reads out the motion. The chair also has the role of maintaining order and the focus of the debate. It is important to establish ground rules for debate to be clear and fair for everyone. It could, for example, be established that an individual may only make one contribution to the debate in order to prevent a few individuals dominating proceedings. It is also important that everyone's right to be heard without interruption and the right to different opinions are respected.
4. The proposer presents arguments for the motion and then the opposer presents arguments against the motion. These are the two key performers in the debate. The success or failure of the debate rests in their hands. If their performances are poorly considered and delivered then the activity will fail to engage the audience.
5. The debate is then opened to the floor. At this point, contributions may be sought immediately from individuals. Alternatively, the audience can be invited to form small groups to discuss the points raised and then decide on a point that the group would like to make to the debate. This enables those that are less confident to make a contribution.
6. Depending upon time and how the debate is progressing, it may then be appropriate to invite further individual contributions from the floor. If the debate has gone well and there is engagement within the room, then it is highly likely that there will be individuals who will be keen to make further contributions.
7. The opposer then sums up arguments against the motion and the proposer sums up arguments for the motion.
8. The speaker re-reads the motion and the audience are then invited to vote for or against it.

Case Study 7.4 provides an example of how this form of debate has been put into practice with a very diverse group of students.

Case Study 7.4 – Student Debate

This case study describes a debate with around 100 students with ages ranging from 18–50 and from very diverse backgrounds. These students were commencing a one year foundation programme at Durham University before progressing onto to study degree programmes across a wide range of disciplines.

A debate was introduced as part of the induction week activities. This would provide the opportunity for all the students to interact together and develop their argumentation skills. The debate proved to be highly successful, students engaged with the activity very positively and enjoyed the opportunity.

The motion proposed was "The United Kingdom has contributed more to the sciences than to the arts" and was sent out as part of a welcome pack prior to the students arriving. An outline of the activity was provided and students were asked to prepare a point to argue for or against the motion before arriving.

On the day of the debate, the chemistry lecturer proposed the motion and a social sciences lecturer opposed (an interesting alternative is the proposer and opposer represent the sides of the argument opposite to their current opinion). Short speeches (ten minutes) were presented that made clear points supported by the evidence and created a stimulating atmosphere within the auditorium.

The audience was then provided with five minutes to discuss in small groups. After which, a spokesperson from each group made a point that was agreed within the group. Discussions were now becoming animated and it was clear that there were plenty of people within the room who wished to continue the debate. The students demonstrated excellent debating skills, making well articulated and thoughtful points. There was also evidence of how opinions had changed during the course of the debate.

The opposer and proposer then briefly summarised and responded to the points made. The vote was then taken by a show of hands that, on this occasion, showed equal numbers both for and against the motion. However, the result of the motion was not actually that important. The real result was that this diverse group of students had the opportunity to engage in a performance that had developed their understanding of science and the arts as well as their debating and argumentation skills.

A follow up to such an activity is then to require the students to produce a written essay on the same motion. They now have a rich body evidence to draw on from the debate that can be approached in a similar way in a written genre.

7.5 Drama Games

With some creative thinking on the part of the teacher, many drama games can be used in a chemistry context. Some activities combine the elements of debate and role play as students are engaged in a scenario and undertake different roles in order to consider the issue (Jones, 1997; Jackson and Walters, 2000; Smythe and Higgins, 2007). Cook (2014) describes a role-playing game exploring the chemistry of plastics. The students are presented with a fictitious scenario where the government has passed a new law to regulate plastic waste. A public hearing is held and a range of interested parties from industry, health and environmental sectors are invited. The students are assigned the different roles in order to debate the issue. The activity is clearly structured with good supporting materials – essential for the success of this complex activity. The teachers involved were extremely positive about their experience and the benefits for the students. The process engaged a diverse group of students and enabled them to develop critical thinking and argumentation skills.

A variation on the formal debate structure are activities such as the "hot air balloon" game. In this activity a number of people are in a hot air balloon that is falling to the ground so someone will have to jump out and make the balloon lighter. Each person makes their case as to why they should stay and then a vote is taken and the person with the least number of votes is out. This continues until the last person remains. This activity can be used in a chemistry context with each person acquiring the character of a famous chemist or they could be a different element and explain why they are more important than the other elements.

Element speed dating can be a lively way to get students interacting and thinking about the structure and properties of different elements. Each student is a different element and spend one minute talking to an element before moving on to the next one. The objective is to find the perfect chemical match, that is to say, another element that they could bond with. This can be extended into other curriculum areas such as displacement reactions or electrochemical cells where students have to find someone with whom there would be a feasible reaction.

7.6 Conclusion

This chapter has focused on how creative approaches to drama and performance can be used to engage a wide diversity of students in chemistry and develop chemical understandings. Key components of successful performance by the teacher have been explored as well as the potential to use drama based activities such as plays, debates and games. Used skilfully and appropriately, these pedagogical strategies have the potential to deepen student appreciation of chemistry, its impact on their lives and how discoveries are made. We should, however, end with a note of caution. Teachers should not forget their role is primarily to teach chemistry and engage students in learning it; moderation in all approaches is often a virtue.

References

Anderson M., (2004), The professional development journeys of drama educators, *Youth Theatre J.*, **18**(1), 1–16.

Anderson C., (2016), *TED Talks: The Official TED Guide to Public Speaking*, Houghton Mifflin Harcourt.

Aubusson P. J. and Fogwill S., (2006), Role play as analogical modelling in science, *Metaphor and Analogy in Science Education*, Dordrecht: Springer, pp. 93–104.

Bailey P. D., (2008), Should 'teacher centred teaching' replace 'student centred learning'?, *Chem. Educ. Res. Pract.*, **9**(1), 70–74.

Cook D. H., (2014), Conflicts in chemistry: The case of plastics, A role-playing game for high school chemistry students, *J. Chem. Educ.*, **91**(10), 1580–1586.

Djerassi C., (2012), *Chemistry in Theatre Insufficiency, Phallacy or Both*, London: Imperial College Press.

Dorion K. R., (2009), Science through drama: A multiple case exploration of the characteristics of drama activities used in secondary science lessons, *Int. J. Sci. Educ.*, **31**(16), 2247–2270.

Duveen J. and Solomon J., (1994), The great evolution trial: Use of role-play in the classroom, *J. Res. Sci. Teach.*, **31**(5), 575–582.

Faraday M., (1865), in Crookes W. (ed.), *A course of six lectures on the chemical history of a Candle; to which is added a lecture on Platinum... delivered during the Christmas Holidays of 1860-1*.

Faraday M., (1960), *Advice to a Lecturer*, Royal institution.

Harvard Project-Zero, (2019), Project Zero, Available at: http://www.pz.harvard.edu/.

Heathcote D., (1991), *Collected Writings on Education and Drama*, Northwestern University Press.

Jackson P. T. and Walters J. P., (2000), Role-playing in analytical chemistry: The alumni speak, *J. Chem. Educ.*, 77(8), 1019.

Jones M. A., (1997), Use of a Classroom Jury Trial To Increase Student Perception of Science as Part of Their Lives, *J. Chem. Educ.*, **74**(5), 537.

Mahaffy P., (2006), Moving chemistry education into 3D: A tetrahedral metaphor for understanding chemistry. Union Carbide Award for Chemical Education, *J. Chem. Educ.*, **83**(1), 49.

Metcalfe R. J. A., Abbott S., Bray P., Exley J. and Wisnia D., (1984), Teaching science through drama: An empirical investigation, *Res. Sci. Technol. Educ.*, **2**(1), 77–81.

Ødegaard M., (2003), Dramatic Science. A Critical Review of Drama in Science Education, *Stud. Sci. Educ.*, **39**(1), 75–101.

Smythe A. M. and Higgins D. A., (2007), (Role) playing politics in an environmental chemistry lecture course, *J. Chem. Educ.*, **84**(2), 241.

Solomon J., (1990), *The Retrial of Galileo (SATIS 16-19, Unit 1*, Hatfield: Association for Science Education, pp. 1–4.
Somers J. W., (1994), *Drama in the Curriculum*, Cassell.
Wagner B. J. and Barnett L. A., (1998), *Educational Drama and Language Arts: What Research Shows*, Portsmouth, NH: Heinemann, p. 231.
Wilhelm J. D. and Edmiston B., (1998), *Imagining to Learn: Inquiry, Ethics, and Integration Through Drama*, Heinemann.

CHAPTER 8

Practical Chemistry

Few would disagree about the importance of practical investigations for chemistry learning and the research evidence supports several positive outcomes (EEF, 2019a, b) such as engaging pupils, developing scientific reasoning skills and a positive impact on attitudes and attainment. It is through practical investigations that students are able to explore and develop an understanding of the nature of matter and chemical interactions. However, practical work has been criticised for its "recipe" card approach and experiments that are predetermined confirmatory evidence of foregrounded knowledge.

In this chapter, we focus on approaches to practical work that provide opportunities for creative thinking and problem solving. We shall explore the scope for escape rooms in chemistry, problem solving practical work and microscale chemistry as creative alternatives to traditional expository experiments.

8.1 Types of Practical Chemistry Instruction

Traditional practical chemistry requires students to follow a predetermined experimental procedure in a set period of time; referred to as "recipe style" or expository (Domin, 1999) laboratory classes. These classes provide the students with the opportunity to experience and exemplify knowledge as well as develop practical and manual dexterity skills. However, these experiences have been criticised for a limited level of learning. Students may be unclear about the aims of the practical or how it applies to chemical concepts (McGarvey, 2004) and are focused on the practicalities of completing the steps correctly. Furthermore, there is limited scope for creativity or contextualisation with experiments simply verifying previously stated knowledge (McDonnell *et al.*, 2007). This format does not allow the student to think about the larger purpose of their investigation (Hofstein and Lunetta, 2004). All too often, over the course of my career, I know that I have been guilty of

Advances in Chemistry Education Series No. 4
Creative Chemists: Strategies for Teaching and Learning
By Simon Rees and Douglas Newton

Published by the Royal Society of Chemistry, www.rsc.org

following a teaching pattern involving explaining a chemistry concept and then the students undertaking an experiment to confirm that I was right! Alternatively, students could undertake the experiment and then generate ideas to explain the data.

Recognition of these limitations has led to the development of pedagogy that is more closely aligned with the nature of science and scientific inquiry. These more open ended and less prescriptive processes have been variously defined as Inquiry, Discovery (or guided inquiry), Process Orientated Guided Inquiry Learning (POGIL) and Problem based learning (Domin, 1999; Eberlein *et al.*, 2008). While there are differences in approach with these different types of pedagogy they emphasise student generated questions and the development of procedures to answer those questions.

The Gatsby International study (Holman, 2017) established ten benchmarks for good practical science. In relation to practical science to promote creativity, key aspects of this are:

- The purpose, outcomes and how these will be achieved through practical work should be clearly defined. A valid purpose of practical work is to promote creative thinking and if this is an established learning outcome then the activity must support this.
- The importance of expert teachers. In this context, this means experts in their understanding of creativity and how to use pedagogy to promote it.
- Frequent and varied practical work. Creative approaches to practical work, as illustrated in this chapter, provide the opportunity to give students more varied practical experiences.
- Students should have the opportunity to undertake independent research projects.
- Assessment that is fit for purpose. If creativity is to be assessed (see Chapter 10), how can this be incorporated appropriately within the activity?

8.2 Escape Rooms

Escape rooms are an example of games based learning (Prensky, 2001) which recognises the established role of games and play in human civilisation as a mechanism for imparting knowledge and social interaction. In a typical escape room experience, players are "locked" in a room, engage in team based games and have a fixed period of time to discover clues, solve puzzles and accomplish tasks in order to achieve a specific goal such as escape from the room. Escape rooms have a strong narrative, and props and scenery are used to create an atmosphere and engage the participants in the activity.

8.2.1 Designing an Escape Room Activity

An escape room should aim to get people in "the zone" where the participants are lost in the moment and are completely focused on the task. Psychologist

Mihaly Csikszentmihalyi called this state "flow" (Csikszentmihalyi *et al.*, 2014) and described it as having the following characteristics:

- intense and focused concentration on the present moment;
- merging of action and awareness;
- a loss of reflective self-consciousness;
- a sense of personal control or agency over the situation or activity;
- losing track of time;
- experience of the activity as intrinsically rewarding.

In addition, an escape room should avoid tasks that are too laborious or too hard. The tasks need to be challenging but not so hard that enthusiasm wilts and the participants lose interest. Rules should be clear and multiple attempts possible with immediate feedback. Instructional safeguards such as hints can be incorporated to keep participants on track. In addition, suitable investment should be made into props and resources to add authenticity such as a piece of aged parchment rather than a laminated card. The games and puzzles should be aligned to learning outcomes and based on sound pedagogical principles. An effective debrief is also important to reflect on the challenges and the learning experience.

The Eduscapes project (Eduscapes, 2019) organises workshops that guide participants through the entire process of developing a successful escape room. These workshops have been successfully trialled in high school settings. Beginning with playing an example escape room and exploring puzzle design, the participants then design and test their own escape room before trying it out on an audience. Engaging in this design process develops creative thinking, problem solving and technical skills. They encourage collaborative learning and an iterative process of trial and improvement. This creates a cycle of productive failure where there are no definitive right answers and promotes risk taking. The escapED project (Clarke *et al.*, 2017) has adopted a similar "learning by designing" approach in higher education. They successfully used the approach to aid other educators in developing their own, live-action games for the purposes of education. Specifically in the chemistry context, Peleg *et al.*, (2019) describe the development and implementation of a chemistry escape room with 1500 high school students. The challenges have a strong emphasis on acid base chemistry. They report high levels of engagement, increased levels of efficacy and effective teamwork. The use of chemistry escape rooms has also been reported in France (Dietrich, 2018) and Malaysia (Nguyen, 2018).

Some practitioners have used the escape room idea to develop puzzle based activities suitable for use within a typical laboratory (Allan (2019a, b) for example – see Case Study 8.1). At the University of Calgary, Canada, the ChemEscape group have developed a "Battle Box" (Figure 8.1). This self-contained puzzle unit is designed to increase students' creative capacity and critical thinking skills. A Battle Box consists of two sides with four puzzles each. The puzzles on each side of a Battle Box may or may not be identical. Two teams of participants, one on each side of the unit, "battle" against

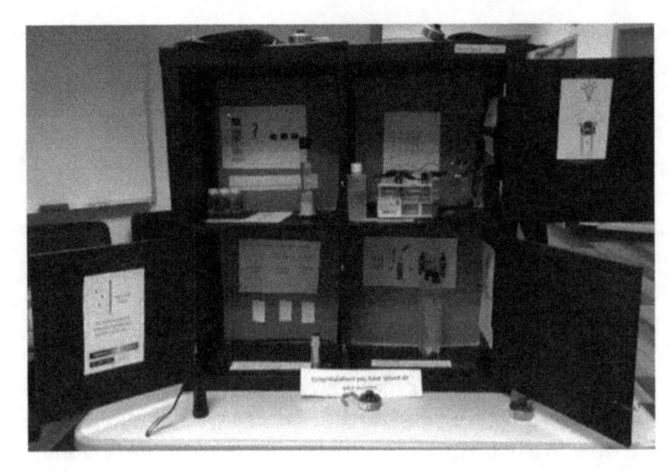

Figure 8.1 The Chemistry Battle Box (image courtesy of Brian Gilbert, University of Calgary).

each other to solve all the puzzles. Starting in the upper-left hand corner of the box (puzzle 1), each puzzle solution opens a lock on the next door, moving in a clockwise manner. The solution of the fourth puzzle corresponds to a lock on a small pencil case, inside of which is a "reward". A puzzle is solved when specific learning objective(s) are met. Ideally, the solution to the fourth puzzle has participants' bringing together all the conceptual knowledge of the previous three puzzles.

The success of opening each lock provides immediate feedback on participants' understanding of the learning objectives. Clues incorporated into the puzzles are structured to lead participants to use specific knowledge. Hints or supplemental information may be given by the facilitator to help participants when knowledge is lacking. Battle Box promotes student-to-student communication and problem solving and puzzles can be designed to use all senses. Allan (2019a, b) describes, in Case Study 8.1, a similar puzzle based approach.

Case Study 8.1 – Escape the Classroom

In a typical escape room game, participants are locked in a room and must solve puzzles as a team to escape. Adrian Allan, chemistry teacher at Dornoch Academy (a remote school in the Highlands of Scotland) developed an alternative format to use in a classroom. The students work as a group to solve various chemistry-based puzzles to open a locked box (Figure 8.2), allowing them to develop problem solving, practical and teamwork skills.

This activity was undertaken as part of a science week with primary and secondary stage pupils aged 10–18 years old. The classroom was set up with five escape room boxes and the class was divided into five groups with each one allocated a box to unlock. The scenario was that the pupils

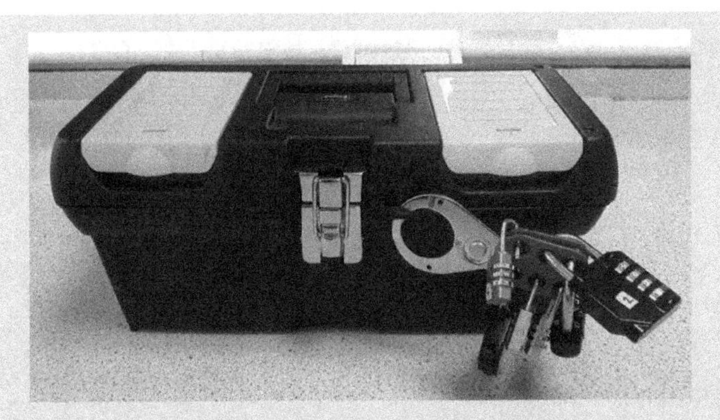

Figure 8.2 The Escape Room box set up (image courtesy of Adrian Allen).

had been infected with a poison and the antidote (which resembled confectionery!) was in the box.

Periodic Table Guess Who? puzzles and chemistry jigsaws needed teamwork and knowledge to solve. The difficulty of the questions in these tasks was tailored to age and stage. Practical and observation skills were developed using conductivity and microscale experiment puzzles. Novel chemicals and materials were introduced using thermochromic paint and UV beads to obtain codes.

The benefits of the activity were that it required a variety of skills the students had learned in science classes to solve the problems, which resulted in a tangible reward at the end. A recap was done at the end of the activity to show the relevance of the chemistry in the tasks they did, such as the use of UV light to detect forged banknotes and the use of thermochromic materials to detect temperature changes. As well as promoting creative thinking and collaborative skills, the students enhanced their knowledge of chemistry and saw the relevance of it in their lives.

There are, therefore, two approaches that can be taken when considering how to incorporate escape rooms into the chemistry curriculum. One is to get together as a group of teachers (Activity 8.1) and undergo the process of designing an escape room that could be used with students. Alternatively, the students can be engaged in the process of designing an escape room themselves. For the time and resource limited teacher such activities may be considered not practical or feasible – a point made by high school teachers in Israel undertaking a CPD programme incorporating escape rooms (Rosenfeld *et al.*, 2019). The creative teacher, however, recognises the value of engaging students in these types of activities. That is not to say, of course, that this type of activity should be the norm or even a frequent experience for the students. Novel activities can quickly lose their value if they are overused. However, to be able to provide students with these activities on a few occasions can lead to memorable experiences and improve engagement. Activity 8.1 is an outline of the process to design an escape room.

Activity 8.1 – The Chemistry Escape Room

This activity is designed to guide you through the process of thinking about designing a chemistry themed escape room activity to address specific curriculum areas.

i) Think of an area of the curriculum that you would like to be the focus for the activity. This may be an area that is more challenging, problematic or uninspiring for the students and, therefore, a good area to target with a more innovative activity.
ii) What are key learning outcomes for this area?
iii) Thinking divergently, how can these be achieved by being presented in the form of puzzles or challenges?
These puzzles may involve a combination of practical work, logical reasoning and problem solving.
iv) What scenario and narrative could be used to frame the activity?
This may be an opportunity to incorporate other interests in your life and cross-curricular links such as travel by setting the scene inside an Egyptian tomb or science fiction and linking with a famous story or film.
v) What reflective activity can be incorporated for participants to consider learning outcomes (both chemistry specific and creative thinking skills such as problem solving, teamwork and collaboration)?

8.3 Microscale Chemistry

Microscale chemistry encompasses the development of a wide range of practical experiments that enable students to observe chemical phenomena on a smaller scale than traditional bench chemistry. Worley *et al.*, (2019), for example, describe the use of microscale chemistry to illustrate the processes of dissolution, ion mobilisation and precipitation during precipitation reactions. The development of these procedures is an example of creative thinking that has several benefits for teaching chemistry. Fewer resources are used which has time, cost, environmental and health and safety benefits. Furthermore, students are able to carefully observe chemical changes. Microscale experiments occur on a smaller scale than traditional experiments but they remain observable macroscopic phenomena. Therefore, it may be more appropriate to regard this as "small scale" chemistry rather than microscopic. A hand lens, or possibly even a microscope, may enhance observations but is certainly not essential for the experiments. The microscale set up enables the student to observe the macroscopic events associated with each of these processes and promotes the development of creative thinking to imagine these processes at the sub-microscopic level.

Microscale chemistry procedures were developed in the USA to address green and waste issues and in South Africa to provide practical work, *via* UNESCO, to countries with limited resources. In the UK, Bob Worley, working with CLEAPSS, an organisation that supports practical science in schools, has developed a wide range of microscale experiments (Worley *et al.*, 2019). These experiments were developed as alternatives to traditional methods that had been in use for over 100 years and addressed several important issues:

- activities that have caused a safety issue such as toxic gas release issues, burns, explosions;
- reduce the cost of expensive equipment by an alternative design or modern materials;
- safety legislation for school practical work;
- speed up practical work so that classroom management is more efficient;
- reduce time and effort in preparation and waste procedures.

These experiments were developed through a process of repeated failures until a successful method was produced. Creative solutions include replacing test tubes with small "puddles" of the reacting solutions on laminated sheets such as when investigating different indicators (Figure 8.3) or the electrolysis of salt solutions such as aqueous copper chloride within a petri dish (Figure 8.4).

Some chemistry teachers expressed objections to the microscale experiments because the procedure was not as required by the examination and

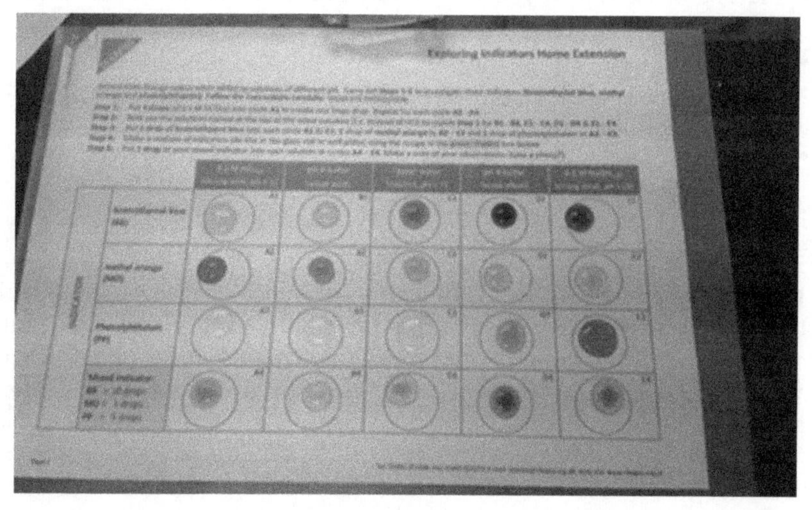

Figure 8.3 pH indicator microscale chemistry (image courtesy of Bob Worley).

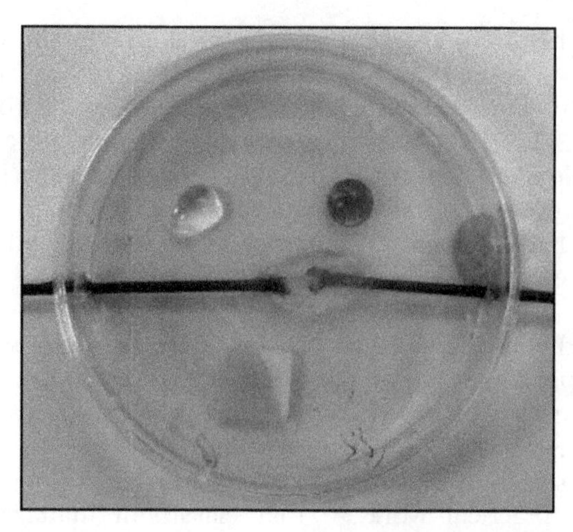

Figure 8.4 Microscale electrolysis of copper chloride solution. Note the bleached blue litmus paper indicating the presence of chlorine (image courtesy of Bob Worley).

students not having the motor skills to undertake the experiments. However, those that have adopted the experiments report several advantages such as:

- laboratory management is easier;
- students are more focused and not side-tracked by other groups;
- mistakes can be quickly and safely corrected;
- the practical sessions are completed in the time available;
- the results are visually pleasing, photographed and inserted into the lab book for future reference;
- the use of webcams, visualisers and USB microscopes allow dramatic demonstrations;
- there is reduced overload of the short-term working memory;
- the practical work focusses more on developing conceptual understanding;
- completely new and interesting experiments have been developed.

8.4 Inquiry Based Learning

In this section, we explore the application of inquiry based learning within practical chemistry. These approaches promote curiosity and questioning, focusing on scientific enquiry and scientific thinking rather than scientific content. Inquiry is unlikely to lead students to discover knowledge unknown to anyone before but it does enable the students to discover new knowledge to them. We will consider evidence and exemplars of good practice to demonstrate how creative practical investigations can promote these areas.

Children are naturally inquisitive and seek to develop a greater understanding of the world around them through their experiences. Formal science education is often criticised as it tends to over emphasise content and facts rather than a method to discover new ideas and knowledge (Soares, 2016). Practical chemistry provides the principle route for many children to undertake first-hand investigative experiences that are both mentally and physically engaging.

Inquiry based learning is grounded in Piaget's epistemological ideas of constructivism (Piaget, 1952) and uses learning cycle (Marek and Cavallo, 1997) methodology of exploration, invention and application. In the exploration stage students obtain data, invention requires deriving the underlying concept and then the concept is applied to other situations (application).

Inquiry based activities have an undetermined outcome and require students to generate their ideas and procedures. There is less direction and students are given more responsibility and a greater sense of ownership (Roth, 2012). Inquiry based learning promotes higher order and creative thinking processes such as hypothesising, analysing, criticising, inventing and evaluating (Raths *et al.*, 1986). Inquiry based learning leads the student to think more like a "scientist" with an emphasis on exploring what experimental data means rather than on chemically correct deductions. Inquiry based learning is more demanding and time consuming than expository activities; it places additional cognitive demands on students by requiring them to simultaneously process new subject matter, unfamiliar laboratory equipment and problem solving tasks (Linn, 1977).

In the UK, at the primary level (up to age 11), there has been increasing emphasis on developing scientific enquiry and promoting open questioning by the students. The Wellcome Trust's Explorify project (Wellcome Trust, 2019), for example, provides resources that promote students to start thinking about everyday phenomena scientifically and to start asking questions. This can then be used as a "launch pad" for the students to start designing investigations that could test the questions they have developed themselves. Such an open ended process reduces the emphasis on single "correct" answers and promotes divergent thinking where more possibilities can be suggested and explored. The skill of the creative teacher is, therefore, to be able to facilitate and guide this process to achieve meaningful outcomes.

Similarly, the Thinking Doing Talking Science project (TDTS, 2019) provides training for primary teachers in the UK to develop creative and challenging science lessons that develop questioning and scientific enquiry. Trials indicated that year 5 (age 9–10) students in schools applying the TDTS approach showed, on average, three months additional progress in science compared to similar students in schools not following the same approach (EEF, 2019a, b). There were also significant increases in positive attitudes towards science and practical work by the students and teachers also reported enjoying teaching science more.

With older students, Lamba (2015) describes an inquiry based curriculum where concepts are introduced inductively *i.e.* derived from the investigations. Students are presented with an initial observation such as that post 1982 North American penny coins react vigorously with acid while pre 1982 coins do not. This leads on to a series of experiments where the students explore the volume of gas produced when a range of metals react with acids. Through guided instruction, the students obtain data which can lead to exploring a range of different concepts such as reactivity, amount of substance and stoichiometry. Lamba (2015) describes a good laboratory method for this experiment that enables the collection of accurate and reliable data. This is important when the emphasis is on interpreting the data to derive the concept. Working at an interdisciplinary level, Rayner *et al.* (2013) successfully developed undergraduate inquiry orientated practicals across Biology, Chemistry and Physics that stimulated a new culture of teaching and learning at Faculty level.

8.5 Problem Based Learning

Problem based learning requires students to apply their conceptual understanding to solve a problem posed by the teacher. This can be contextualised to a real world problem. Students may be provided with guidance over materials and steered towards a successful solution to the problem. They may already have the required knowledge to solve the problem or have to identify and acquire relevant information and procedures. The key feature is that the learning process is initiated by the problem (Poë, 2015) and students, normally working in groups of four or five follow a series of tasks towards a solution. These tasks require the students to analyse the problem and establish what knowledge and skills are required to address it. This requires both evaluative and creative thinking. Tasks are assigned within the group and new knowledge is synthesised and reviewed to construct and report on a solution to the problem.

An important aspect of this pedagogy is the emphasis on students determining what it is they need to find out in order to solve the problem. This may initially be viewed as a daunting task for students who are used to receiving new knowledge and then applying it in particular contexts. Students benefit from being guided towards this way of thinking over a period time and appreciating the problem solving skills they are developing. McDonnell *et al.*, (2007) successfully developed PBL mini projects in place of traditional recipe style practicals for undergraduate chemistry students. They reported that students required more support initially until they became more familiar with the process and it led to improved engagement.

One criticism of PBL is the time it takes and this is not possible when so much content needs to be covered. This can only be addressed through curriculum design where equipping students with understanding of scientific method and problem solving skills is valued as well as specific chemical knowledge. There are also many hybrids of the PBL approach that can incorporate elements of the process without unduly occupying curriculum time. In addition, digital resources such as virtual experiments can provide excellent opportunities to familiarise students with equipment and procedures so that contact time can be used more effectively.

8.6 Virtual Experimentation

The use of digital technologies has led to the development of a range virtual experiments and simulations to assist students. The Royal Society of Chemistry virtual experiments, for example, enable students to undertake organic synthesis and volumetric analysis using a context based approach (RSC, 2019). These resources provide the opportunity for students to familiarise themselves with unfamiliar experimental equipment and procedures so that they can engage with practical work more effectively and think more about the chemistry concepts involved. They also enable opportunities to reinforce understanding and apply learning to new contexts in a way that would not normally be possible. Similarly, resources such as the Phet (2019) simulations enable students to investigate the effect of changing variables in a system far more quickly than would be possible in the laboratory. It is important that curricula are designed where these resources are an augmentation that enables more effective practical work rather than a replacement.

8.7 Science Technicians

A very important consideration in developing creative practical chemistry is engaging and valuing the contribution of science technicians in this process. Technicians can explore different methods for undertaking experiments, devise new pieces of equipment and ensure activities are organised as effectively as possible. Engaging in the development of new activities is an important part of a technician's professional development and can be used as evidence to obtain professional status such as the Royal Society of Chemistry Registered Science Technician award.

Activity 8.2 – There Are Many Ways to Run a Practical

This activity is designed to help you think creatively about how to deliver a practical with your classes.

i) Think of a practical that is currently delivered in an expository or "recipe book" style.
ii) Using divergent thinking, think of as many different ways as possible to deliver the same practical. This could range from presenting the students with a sorting activity where the instructions are jumbled up and have to be sorted into the right order to an extended inquiry based learning activity involving the students in developing their own questions and procedures.
iii) Now, using evaluative thinking, reflect on the different strategies and the different learning outcomes and creative capabilities they address.

iv) Choose three of these strategies and develop individual lesson plans to be able to deliver the same practical in three different ways. If the practical is inquiry or problem based then consider how these thinking skills can be introduced and developed over a period of time.

You now have a choice of strategies for delivering the same curriculum area that you can choose to apply depending upon the teaching context and the experience that you would like the students to have.

8.8 Summary

In this chapter, we have explored the wide variety of ways that practical chemistry can be undertaken to engage students and promote creative thinking. Whether it is presenting content in novel ways such as escape rooms or designing inquiry based investigations, there can be no doubt that a well designed and structured curriculum will make full use of this range of activities to provide a stimulating and accessible experience for all students. This also helps to maintain teacher enthusiasm as the potential for new and varied pedagogies are explored. As chemistry teachers, we have the good fortune of teaching a subject with content that can be exploited to develop interesting and stimulating practical opportunities.

References

Allan A., (2019a), Escape the Classroom, Available at https://eic.rsc.org/ideas/escape-the-classroom/3009832.article.

Allan A., (2019b), Escape the Classroom – again! Available at: https://edu.rsc.org/ideas/escape-the-classroom-again/4010759.article.

Clarke S., Peel D. J., Arnab S., Morini L., Keegan H. and Wood O., (2017), escapED: a framework for creating educational escape rooms and Interactive Games For Higher/Further Education, *Int. J. Serious Games*, **4**(3), 73–86.

Csikszentmihalyi M., Abuhamdeh S. and Nakamura J., (2014), Flow, *Flow and the Foundations of Positive Psychology*, Dordrecht: Springer, pp. 227–238.

Dietrich N., (2018), Escape Classroom: The Leblanc Process—An Educational "Escape Game", *J. Chem. Educ.*, **95**(6), 996–999.

Domin D. S., (1999), A review of laboratory instruction styles, *J. Chem. Educ.*, **76**(4), 543.

Eberlein T., Kampmeier J., Minderhout V., Moog R. S., Platt T., Varma-Nelson P. and White H. B., (2008), Pedagogies of engagement in science: a comparison of PBL, POGIL, and PLTL, *Biochem. Mol. Biol. Educ.*, **36**, 262–273.

Eduscapes, (2019), Available at: http://eduscapes.playthinklearn.net/.

Education Endowment Foundation (EEF), (2019a), Improving Secondary Science Guidance Report, Available at: https://educationendowmentfoundation.org.uk/public/files/Publications/Science/EEF_improving_secondary_science.pdf.

Education Endowment Foundation (EEF), (2019b), *Thinking Doing Talking Science*, Available at: https://educationendowmentfoundation.org.uk/projects-and-evaluation/projects/thinking-doing-talking-science.

Hofstein A. and Lunetta V. N., (2004), The laboratory in science education: Foundations for the twenty-first century, *Sci. Educ.*, **88**(1), 28–54.

Holman J., (2017), *Good Practical Science*, London: Gatsby Foundation.

Lamba R. S., (2015), Chemistry Education: Best Practices, Opportunities and Trends. Editor (s): Javier García-Martínez, Elena Serrano-Torregrosa.

Linn M. C., (1977), Scientific reasoning: Influences on task performance and response categorization, *Sci. Educ.*, **61**(3), 357–369.

Marek E. A. and Cavallo A. M., (1997), *The Learning Cycle: Elementary School Science and Beyond*, Heinemann.

McDonnell C., O'Connor C. and Seery M. K., (2007), Developing practical chemistry skills by means of student-driven problem based learning mini-projects, *Chem. Educ. Res. Pract.*, **8**(2), 130–139.

McGarvey D. J., (2004), Experimenting with undergraduate practicals, *Univ. Chem. Educ.*, **8**, 58–65.

Nguyen T., (2018), Chemistry education moves from classroom to escape room, *Chemical and Engineering news*, https://cen.acs.org/education/k–12-education/Chemistry-education-moves-classroom-escape/96/i19.

Peleg R., Yayon M., Katchevich D., Moria-Shipony M. and Blonder R., (2019), A Lab-Based Chemical Escape Room: Educational, Mobile, and Fun! *J. Chem. Educ.*, **96**(5), 955–960.

Phet, (2019), Interactive Simulations for Science and Math, https://phet.colorado.edu/.

Piaget J., (1952), *The Origins of Intelligence in Children*, New York: International Universities Press.

Poë J. C., (2015), Active Learning Pedagogies for the Future of Global Chemistry Education, in Garcia-Martinez J. and Serrano-Torregrosa E. (ed.), *Chemistry Education: Best Practices, Opportunities and Trends.*

Prensky M., (2001), Fun, play and games: What makes games engaging, *Digital game-based Learn.*, **5**(1), 5–31.

Raths L. E., Wassermann S., Jonas A. and Rothstein A., (1986), Teaching for Thinking: Theories, *Strategies, and Activities for the Classroom, Teachers College*, New York: Columbia University.

Rayner G., Charlton-Robb K., Thompson C. and Hughes T., (2013), Interdisciplinary collaboration to integrate inquiry-oriented learning in undergraduate science practicals, *Int. J. Innov. Sci. Math. Educ.*, **21**(5), 1–11.

Rosenfeld S., Yayon M., Halevi R. and Blonder R., (2019), Teachers as Makers in Chemistry Education: an Exploratory Study, *Int. J. Sci. Math. Educ.*, 1–24.

Roth W. M., (2012), *Authentic School Science: Knowing and Learning in Open-inquiry Science Laboratories*, Springer Science & Business Media, vol. 1.

RSC (Royal Society of Chemistry), (2019), Screen Experiments, Available at: http://www.rsc.org/learn-chemistry/collections/experimentation/screen-experiments.

Soares L., (2016), Sciencing: Creative, Scientific Learning in the Constructivist Classroom, *Interplay of Creativity and Giftedness in Science*, Brill Sense, pp. 127–151.

TDTS, (2019), Thinking Doing Talking Science, https://tdts.org.uk/.

Welcome Trust, (2019), Explority, https://explorify.wellcome.ac.uk/.

Worley B., Villa E. M., Gunn J. M. and Mattson B., (2019), Visualizing Dissolution, Ion Mobility, and Precipitation through a Low-Cost, Rapid-Reaction Activity Introducing Microscale Precipitation Chemistry, *J. Chem. Educ.*, **96**(5), 951–954.

CHAPTER 9

The Language of Chemistry

In this chapter, we explore the central importance of the language of chemistry in creative thinking. We discuss the research evidence highlighting the challenges it presents such as: specific technical words and symbols, words used in a variety of contexts or everyday words that have different meanings in chemistry. These challenges are explored through the lens of modelling linguistic demand in multiple dimensions in order to promote the development of "language consciousness" amongst chemistry educators. Awareness of the importance of these issues leads to the creative development of pedagogical strategies to support students in developing their understanding of chemical language. Vygotsky (1962) made the point very succinctly when he wrote *"Language development and conceptual development are inextricably linked. Thought requires language, language requires thought."* Not everything that must be learned in chemistry always lends itself readily to creative teaching and learning – the language of chemistry can be, at times, an instance. Learning the language of chemistry and what words mean takes time, but making them meaningful is something the teacher should not neglect. This is where imagination can be an asset to contribute to the quality of creative teaching in all areas of chemistry.

9.1 Scientific Language

Returning to Michael Faraday's lectures "On the Chemical History of a Candle" (Faraday, 1865), introduced in Chapter 7, we are reminded of the importance of language for communication and understanding.

> *"And, before proceeding, let me say this also: that, though our subject be so great... I mean to pass away from all those who are seniors among us. I claim the privilege of speaking to juveniles as a juvenile myself... though I stand here with the knowledge of having the words I utter given to the world, yet that*

Advances in Chemistry Education Series No. 4
Creative Chemists: Strategies for Teaching and Learning
By Simon Rees and Douglas Newton

Published by the Royal Society of Chemistry, www.rsc.org

shall not deter me from speaking in the same familiar way to those whom I esteem nearest to me on this occasion."

In so doing, Faraday demonstrates his great understanding as a science communicator by recognising the importance of the language that he will use to engage his audience. Language is the medium by which we communicate ideas and develop understanding of chemistry. However, the language of science has been described as "monolithic castle of impenetrable speech" (Montgomery, 1996) that is not readily acquired in general communication. Faraday considered carefully the words and language he would use to ensure it is accessible to the non-scientific audience. Accessibility, however, is not easy to achieve, nor is it always in the longer term interests of a learner of chemistry. Understanding the subject of chemistry ultimately means developing an understanding of, and a fluency in the use of, its subject specific language. As Postman and Weingartner (1971, p. 103) state: *"the key to understanding a "subject" is to understand its language...what we call a subject is its language. A "discipline" is a way of knowing, and whatever is known is inseparable from the symbols (mostly words) in which the knowing is codified."*

In order to think creatively or engage in creative discussions with others it is essential to have sufficient fluency in the language of chemistry. This is required to develop conceptual understanding and participate in original problem solving, providing solutions that are both original and useful. Fredericks (2005) highlighted the importance of fluency for an individual to undertake divergent thinking and produce plausible ideas. Limitations in language knowledge can limit the capacity to generate a variety of ideas.

Conceptual development is inextricably linked to language development (Vygotsky, 1962), and as Byrne *et al.*, (1994) state, "*difficulty with language causes difficulty with reasoning*". In chemistry learning specifically, Nyachwaya (2016), for example, demonstrated links between poor conceptual understanding of acid–base chemistry and poor language fluency.

Lee and Fradd (1998) and Lee (2005) demonstrated how a lack of linguistic skills and unfamiliarity with scientific language was demotivating for students to ask questions and undertake investigations. Engagement with these activities, however, are an essential part of inquiry based learning, problem solving and creative thinking. Students are required to interpret and move between different representations such as verbal, written, symbolic and diagrammatic and to integrate these. Therefore, if we are to design teaching activities to encourage creative thinking, we must also support students to become fluent in subject specific language. Wellington and Osborne (2001) recognise the central importance of language in learning science with the following beliefs:

"– Learning the language of science is a major part (if not the major part) of science education. Every science lesson is a language lesson.

– Language is a major barrier (if not the major barrier) to most pupils in learning science." (p. 2)

It is particularly important to engage students who are diverse in terms of culture, language and prior knowledge. Bilingual students' achievements in science, for example, are, on average, lower than those of their monolingual peers (Ünsal *et al.*, 2018). Bilingual students are faced with the greater challenge of not only learning the language of instruction but also the language of chemistry. Hence, enhancing the understanding of chemical language is of paramount importance.

9.2 The Significance of Language in Chemistry Learning

A number of authors have considered the importance of language learning in chemistry in recent years (Laszlo, 2013; Taber, 2015; Markic and Childs 2016 and Rees *et al.*, 2018a). Students often find chemistry difficult (Johnstone, 1991) particularly at introductory levels. Challenges include: developing formal explanations of phenomena at the macroscopic level; the requirement to connect macro- and sub-micro levels (Johnstone's "multi-level" thinking, 1991); the requirement for abstract thought with explanations involving "particles" at a sub-microscopic level which behave differently from macro-scale particles such as sand grains, and the potential for information overload of working memory (Johnstone, 1991).

9.2.1 The Language of Johnstone's Triplet

Johnstone (1991) developed a view that chemistry learning occurs on three levels: macroscopic, that is what can be seen, touched and smelled; sub-microscopic, that is atoms, molecules, ions and structures; and symbolic meaning, that is representations of formulae, equations, mathematical expressions and graphs. Inspired by a geologist's diagram describing mineral composition, Johnstone arranged these levels at the corners of an equilateral triangle (Figure 9.1) to indicate equal, complementary significance. Teaching occurs "within" the triangle, under the assumption that all levels are equally well-understood. During chemistry learning, novice students must move between these three levels, often without notice or explanation. This introduces too much complexity for a novice chemist. A successful learner needs to develop competence in, and then confidently inter-relates these three aspects.

This model can be used to consider the vocabulary required to explain a given chemical phenomenon. Activity 9.1 can be undertaken with students and teachers and reflects on the complexity of the language needed to interpret everyday chemical phenomena and the demands placed on students. These are only increased as we progress into more specialised and less familiar territory. This activity also highlights how, when talking about the language of chemistry, we are talking about much more than just words.

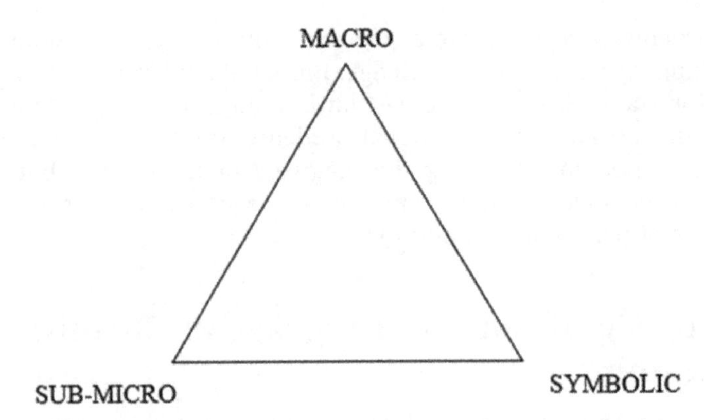

Figure 9.1 The chemistry learning triplet.
Reproduced from Johnstone, 1991 with permission from John Wiley and Sons, Copyright 1991.

Language is the mechanism for communicating knowledge and ideas whether it is through reading, writing, listening or speaking. Each requires different aspects of language use and in chemistry, this language is multifaceted. Markic and Childs (2016) refer to this chemical language as "Chemish" which encompasses words, symbols, equations, diagrams, graphs and mathematical expressions.

Activity 9.1 – Multiple Levels of Language

This activity is designed to help students and teachers think about the range of language required to explain a chemical phenomenon.

i) For any given chemical phenomenon (such as a burning candle) use divergent thinking to write down as many words or symbols that you can think of that are required to describe and explain the phenomenon.
ii) Now draw Johnstone's triangle on a large sheet of paper and assign the words and symbols to the three different levels as shown, for example, for a burning candle (Figure 9.2).
iii) Questions to consider:
- What do you notice? Are there more words and symbols in one level than another?
- What role do the different words or symbols have in developing understanding?
- Are some words more accessible than others?
- How well do students understand the connections between these words?

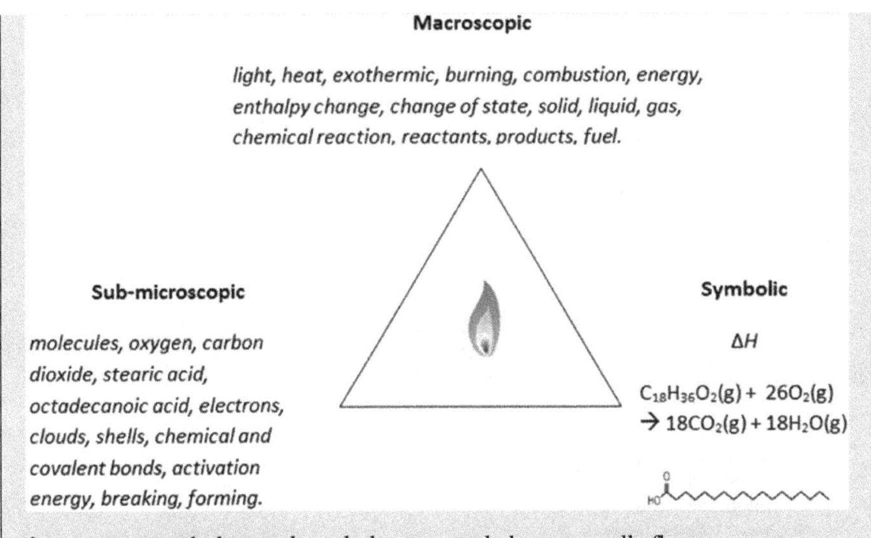

Figure 9.2 Vocabulary and symbols generated about a candle flame. Reproduced from Rees, Kind & Newton, 2019 with permission from John Wiley and Sons, © 2019 Wiley-VCH Verlag GmbH & Co. KGaA, Weinheim.

- What shared meanings are there between the teacher and students?

Taber (2013) revisited the triplet to address two confusions associated with Johnstone's model. Firstly, the macroscopic level in terms of phenomenological and conceptual frameworks related to these phenomena. Secondly, the symbolic level and how this fits as a representational level with the macro and sub-microscopic levels. He argued that conceptual demand is high at the macroscopic corner as students deal with abstract notions relating to substances with unfamiliar names and classifications, for example, *alkali metals, acids* and *reducing agents*. He highlighted the role of specialised language in chemistry and how macroscopic concepts, such as *solution, element* and *reversible reaction* or microscopic, including *electron, orbital, hydrated copper ion* need to be represented for a novice to think about them and communicate understanding to others. He suggested that the symbolic level should not be regarded as discrete in its own right but as a conduit for the representation and communication of chemical concepts (Figure 9.3).

9.2.2 The Words of Chemistry

Learning the words of chemistry presents several challenges. There are words that have an everyday meaning but then a specific and different meaning in chemistry. There are conceptual words such as *energy* or *equilibria,* or naming words, such as *element, compound* and *base,* that have been recruited by chemistry. There are also many new words that are often

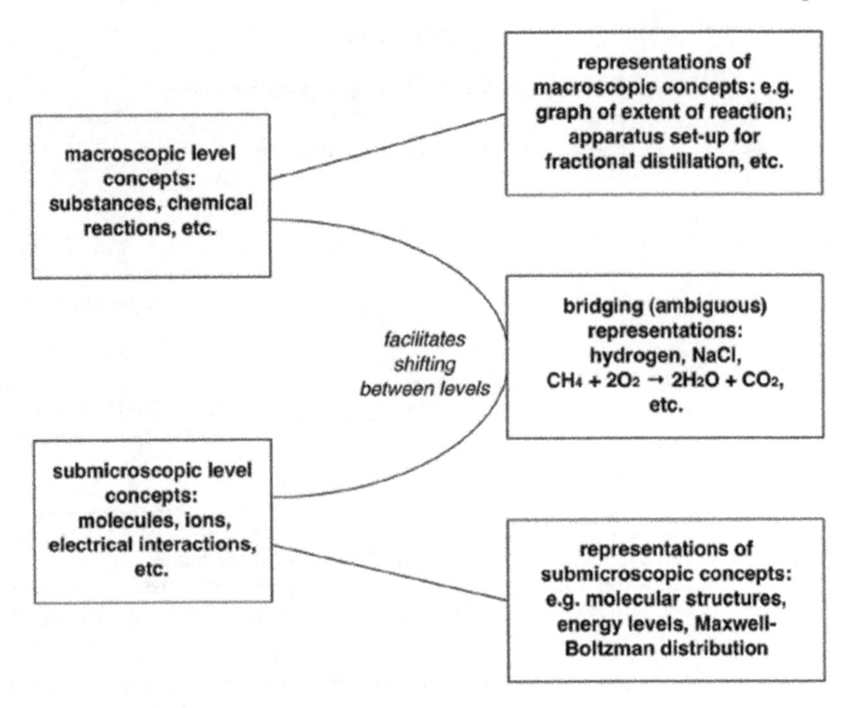

Figure 9.3 The symbolic domain for supporting the development of explanations relating the macroscopic and sub-microscopic levels.
Reproduced from Taber, 2013 with permission from the Royal Society of Chemistry.

introduced in unfamiliar contexts, such as *proton or oxidation*. Furthermore, there are those words, such as *evaluate, modify, constituent* and *replicate,* which are referred to as "the language of secondary education" (Wellington and Osborne, 2001, p. 5). Learners can be unfamiliar with these words as they rarely come across them outside the science classroom.

Eiss (1961) recognised difficulties caused by words with different meanings in separate sciences, such as *weight* in biology and physics. Gardner (1972) carried out an extensive study in this area. He tested over 7000 Australian 11–16 year old students' understandings of 600 words considered by science teachers to be essential or valuable in school science. The words selected were non-technical and unlikely to be taught explicitly, for example, *abundant, affect, conception, partial* and *phenomenon*. He found words frequently used in the science classroom were inaccessible to students, including *consecutive, spontaneous, standard* and *stimulate*. This highlights students' difficulties with "normal" English words used in scientific contexts.

9.2.3 The Importance of Context

Cassels and Johnstone (1980) identified a limitation of Gardner's work, noting that their results were based on one question testing each word.

These authors recognised that meaning is determined by context. They undertook a two-year study in the UK with 25 000 secondary 11 to 17 year olds presenting them with words in multiple contexts. One question type used a one world synonym without context. For example:
Abundant can mean:

a) exact
b) perfect
c) scarce
d) plentiful

The word was also used in a scientific context. For example:
There was an abundant supply of gas to the reaction chemicals.
This means that

a) There was a shortage of gas.
b) The supply of gas was just enough for the reaction.
c) The gas was not suitable for the reaction.
d) There was plenty of gas for the reacting chemicals.

Or the word was used in a non-science context. For example:
Apples were abundant last year.
This means that

a) They were larger than normal.
b) There was a poor supply of them.
c) They were ready for picking earlier.
d) There were plenty of them.

This example could be used to explore your students' understanding of words in a scientific context.

- Do fewer students choose the correct option in a scientific rather than an everyday context?
- Why is the sentence in a scientific context more demanding?
- Can you think of some other examples of words you use when teaching that may be more difficult in a scientific context?

Cassels and Johnstone (1980) established similarities between Australian and UK students' understandings. They reported that a word used in a scientific context is harder to understand than the same word in a non-scientific context. They highlighted word combinations that result in an expression with a difficult meaning. For example, students correctly defined the word *invert* when referring to an egg timer, but only 50% were able to complete the statement "to invert an object means" with the phrase "to turn the object upside down". The egg timer context provided a clue about the meaning of the key word. Similarly, the word *external* when used with *TV aerial* was easy to understand but proved difficult when linked with *skeleton*. Cassels and Johnstone (1980) argue that moving to a scientific context requires students to interpret a completely new context to find meaning for a word. Cassels and Johnstone (1983) discuss the teacher role in seeking connections between new and existing vocabulary and the importance of linking new information to existing relevant concepts. They draw on Ausubel (1963) who suggested that meaningful learning occurs only if new information is linked to existing relevant concepts.

Words do not have universal meanings applicable to every situation but a word or utterance gets its meaning as a result of its use in a particular context. Cassels and Johnstone (1980) demonstrate the importance of word context enabling a student to deduce meaning in science. Cassels and Johnstone (1985) refined their work, focusing on 95 words reported as especially problematic in their previous study. This list included words from "non-scientific" English applied in a scientific context. These authors designed multiple choice questions to test understanding of these words in four formats in 30 000 11–18 year old respondents. Synonym questions appeared to generate the most difficulties as students were required to know the meaning of the word without any contextual cues. Other question styles placed the word in context which may have carried sufficient information to give cues. Words identified as "weak" or "very weak" in terms of understanding are shown in Table 9.1. This list includes words where the opposite meaning was selected, such as *negligible* meaning *a lot*.

In a smaller scale study, Pickersgill and Lock (1991) investigated the use of 30 non-technical words taken from Cassels and Johnstone (1980). About 200 students completed an assessment with questions in four formats. Replicating previous findings, synonym questions proved to be least understood. Pickersgill and Lock (1991) identified instances of pupils taking the opposite meaning to that which was intended as well as choosing words that sounded or looked similar, such as *retract* and *contract*.

Further studies have continued to highlight difficulties faced by students interpreting words in a scientific context (Marshall *et al.*, 1991; Tao, 1994; Farell and Ventura, 1998; Childs and O'Farrell, 2003; Ali and Ismail, 2006). Oyoo (2017), in a study of over 1000 South African secondary school students identified the non-technical words *sensitive, spontaneous, retard, contract* and *convention* as causing most difficulty. Investigating further, Oyoo (2017) interviewed students to explore their understanding of these difficult words. Students typically attributed everyday meaning to the word such as, in the case of *sensitive*, a sensitive person is fragile and should be handled with care. Evidence for lack of precision in student understanding was apparent when choices were swayed by the context. Cassels and Johnstone (1985) express concern:

> *"This can have very serious consequences for concept development. It may be in language that the origins of alternative frameworks lie. Loose language must give rise to loose reasoning and strange conclusions, particularly if opposites emerge"* (p. 14).

Table 9.1 Problematic words used in science identified in Cassels and Johnstone (1985) and Pickersgill and Lock (1991) * indicates the same words identified in both studies.

Cassels and Johnstone (1985)	Pickersgill and Lock (1991)
Abundant*	Abundant*
Contract*	Contract*
Spontaneous*	Spontaneous*
Converse*	Converse*
Adjacent*	Adjacent*
Valid*	Valid*
Incident*	Incident*
Negligible*	Negligible*
Emit*	Emit*
Linear*	Linear*
Random	Liberate
Contrast	Factor
Composition	Concept
Complex	Tabulate
Exert	Conception
Component	Disintegrate
Sequence	Stimulate
Relevant	Retard
	Convention
	Diversity

From the teacher's perspective, lapses into inconsistent and imprecise use of language can affect student understanding significantly. From the students' perspective, developing precise and appropriate language use enables clarity of conceptual understanding. Taber and Coll (2002) make a similar point in relation to discussions that move between the macroscopic and microscopic facets of bonding.

In some instances, the scientific meaning of a word such as *weak* and *strong* (in relation to acids and bases) is different to its everyday meaning. Jasien (2010); Snow (2010); Song and Carheden (2014) recorded difficulties with understanding these words in a scientific context. Rees *et al.*, (2018a) reported how these everyday meanings persisted amongst some students from non-traditional backgrounds for words such as *salt, solution* and *neutral* despite specific language focused instruction. For other words such as *reduction* and *weak*, however, the students showed substantial improvement in their understanding after instruction.

Cink and Song (2016) make similar findings with students from diverse ethno-linguistic backgrounds. They argue that the difficulty with appropriating the scientific meaning was that the everyday meanings were deeply rooted in their discursive identities, before learning the scientific meanings. Those students who did successfully acquire the scientific meanings had multiple opportunities to use the words in a scientific context.

Furthermore, previous studies (Marshall *et al.*, 1991; Pickersgill and Lock 1991; Johnstone and Selepeng, 2001) have reported instances of students confusing the meaning of similar words such as *retract* and *contract, conversion* and *convention*, and *negligent* and *negligible*. Vladušić *et al.*, (2016), in a study of Croatian undergraduate and graduate chemistry students, reported instances of students confusing *težište*, which means the "centre of mass in relation to electron distribution in bonding" with *težnja*, which means "aspiration". Rees *et al.*, (2018a, b) describe instances of students confusing similar words such as *electronegative, negative* and *electron density* when explaining chemical phenomena. These results highlight the importance of introducing scientific vocabulary from an early age and providing multiple opportunities for students to use the words in context.

9.2.4 Inconsistent Textbook Language

Textbook language must also adopt precise and appropriate forms. Pekdag and Azizoglu (2013), for example, studied semantic mistakes in chemistry textbooks from the USA, France and Turkey relating to "amount of substance". They highlighted interchangeable and incorrect uses of the *mole* and *amount of substance* and usage of *element, compound, atom* and *molecule* in ways that lead to students' confusion at macroscopic, microscopic and symbolic levels. Consider the two forms of the following question:

1. *How many moles are there in 6 g of carbon?*
2. *What is the amount of substance, in moles, in 6 g of the element carbon, C?*

The first form of the question may commonly be heard in chemistry classrooms but is semantically incorrect. According to the SI definition of the mole, if the amount of substance is to be expressed by associating it with physical quantities then the macroscopic form of the substance (element) should be expressed (Pekdag and Azizoglu, 2013) *e.g.* 6 g of the element carbon. Furthermore, in sentence 2, *mole* is now correctly contextualised as the *unit* for amount of substance. Mole becomes subsidiary to the concept of amount of substance. The phrase "in moles" can be omitted completely from this sentence and it remains semantically correct. Sentence 2 places *the mole* into a similar context to other units of measurement. For example, it is more common to say "how tall are you?" than "how many metres are you" or "what is the angle in this triangle?" compared to "how many degrees is this angle?" These terms are the unit of measurement for length or angle just as the mole is the unit of measurement for amount of substance and the unit is not pre-eminent in the sentence. Echoing the thoughts of Cassels and Johnstone (1985), Pekdag and Azizoglu (2013) state:

> *"these mistakes are not only capable of obstructing a student's scientific understanding and learning of the quantity of amount of substance and its unit the mole but also have the potential of creating misconceptions as well."* (p. 123)

This can affect students' ability to operate at the macroscopic, submicroscopic or symbolic level. Students may confuse verbal expressions using terms incorrectly from these levels. This imprecise use of language goes beyond traditional textbooks, it can also be observed in other resources such as websites written with little awareness of the challenges that chemistry language can present. Good science is dependent on precision and in the same way; good science teaching is dependent on linguistic precision. It is apparent from this body of evidence that an important part of a creative chemistry teacher's repertoire is to have a conscious awareness and understanding of the challenges that the language of chemistry can present – what we term "chemistry language consciousness". This awareness then leads on to understanding how to apply pedagogical strategies to address these issues.

9.3 Scientific Literacy

The importance of scientific literacy has been recognised globally and has been assessed repeatedly over the past 20 years (OECD, 2015) by the Programme for Individual Student Assessment (PISA). Individuals must be scientifically literate to engage meaningfully with many aspects of a modern, technological society. Scientifically literate people are required to develop creative solutions to key world problems such as energy supply and clean water.

The Organisation for Economic Cooperation and Development (OECD) define scientific literacy as:

> "*...the ability to engage with science-related issues, and with the ideas of science, as a reflective citizen.*
>
> *A scientifically literate person, therefore, is willing to engage in reasoned discourse about science and technology which requires the competencies to:*
>
> 1. Explain phenomena scientifically
>   *Recognise, offer and evaluate explanations for a range of natural and technological phenomena.*
> 2. Evaluate and design scientific enquiry
>   *Describe and appraise scientific investigations and propose ways of addressing questions scientifically.*
> 3. Interpret data and evidence scientifically
>   *Analyse and evaluate data, claims and arguments in a variety of representations and draw appropriate scientific conclusions.*" (OECD, 2015, p. 7)

Creative thinking and language are closely linked throughout this definition with an emphasis on *reasoned discourse, explaining, describing* and *evaluating*. A person requires the appropriate language in order to be able to engage in imaginative and creative thought about scientific issues, their implications and solutions.

9.4 Lexical Quality Hypothesis

Linguistic research has developed the concept of the "lexical quality hypothesis" to describe how well a student knows a word, that is to say, their depth of understanding (Perfetti and Hart, 2002). It is estimated that a learner requires a vocabulary of between 50–60 000 words in order to thrive academically. The lexical quality hypothesis argues that, while the number of words is important, it is more important to consider how well these words are understood – the quality of word knowledge rather the quantity of words known. This requires the learner to develop an understanding of the multifaceted nature of a word including its pronunciation, spelling, multiple meanings, word families and so on. In order to achieve this, as much exposure to the word as possible is required. Learning complex academic words is a slow process requiring multiple exposures to achieve word depth. Beck *et al.*, (2002) indicated that learners require anything between four and ten exposures to a new word before it is likely to be established in the long term memory. How can we use creative thinking to develop imaginative opportunities to improve word depth? How carefully and precisely do we use these words to ensure we are developing depth within clear rather than murky waters?

9.5 Word Classification

Wellington and Osborne (2001) classified science words into different categories as an aid to understanding why they may present more or less difficulty to learners. This enables teachers to become more language aware. The first category, Level 1, consists of "naming words" which includes tangible objects and entities such as *burette* or *sodium*. In science more generally, these words may be synonyms of everyday words that may be familiar to learners *e.g. windpipe* and *trachea* and, therefore, it is a process of giving familiar objects new names. Within chemistry, however, there are fewer of these familiar objects that are given new names. More often, such words (*e.g. burette, conical flask*) are unfamiliar to learners. This is one reason why the language of chemistry can be more inaccessible than in other subjects. Level 2 contains "process words" such as *evaporation* and *crystallisation*. Understanding of these processes can be acquired through demonstration such as boiling a salt solution. Other processes such as electron transfer, however, cannot be demonstrated and require greater creative thought to comprehend. Level 3 contains concept words such as *pressure* and *energy* that present high levels of abstraction and require understanding of many other scientific words. For example, electronegativity requires understanding of electronic structures and covalent bonding.

9.6 Linguistic Demand in Multiple Dimensions

Classifications such as these are useful to appreciate the different types of words and their role in conveying understanding. However, they do not convey a sense of the particular challenges that different words can present. These challenges extend beyond simply understanding the correct meaning of a word but also incorporating words into oral and written explanations. Based on the research evidence, Rees *et al.*, (2019) have developed a model to represent linguistic demand in four different dimensions that we expand on here.

(i) Interpretive

The first dimension of linguistic demand relates to the **interpretive** value of a word. In Roald Dahl's children's story "The BFG" (The Big Friendly Giant (Dahl, 2007)) Sophie, a young girl, befriends a giant whose language includes unique words in unfamiliar contexts. *"What I mean and what I say is two different things," the BFG announced rather grandly. Now let me introduce you to the snozzcumber."* No one, before reading this story, would have any understanding of what "snozzcumber" means (or "gobblefunk", "slopgroggle", and "fizzwiggler" to name a few more), but this inventive word play engages the reader in the story. In the same way, chemistry has its own "snozzcumbers", new words, presented in unfamiliar contexts. The contrast with The BFG is that for many learners this is not an enjoyable experience but becomes a source of frustration and a barrier to accessing the subject. As unfamiliar as snozzcumber is, it has the potential to be interpreted and reveal its meaning. The word contains "clues" from which meaning can be inferred. "Cumber" may be linked to the vegetable known as a "cucumber", while "snozz" may equate to "snot", the colloquial English term for nasal mucus. Hence, a snozzcumber could, and does indeed, mean a revolting, slimy cucumber-like vegetable. The word snozzcumber, therefore, presents low demand in the interpretive dimension because its meaning can be inferred. Words presented in a scientific context are more difficult to understand. Text has a high degree of informational density and abstraction. However, if the meaning of the words can be inferred or interpreted then the language is more accessible.

The interpretive value of words in science was discussed by Sutton (1992). He argues there are two ways of using language in science teaching; namely, *interpretive* and *labelling*. A labelling approach is definitive, assumes meaning is understood and implies there is only one way to "see" events. An interpretive approach is exploratory, recognises there is room over how an idea can be expressed and consciously uses language to help people see a topic in new ways. The former offers little room for creative thought – it is equivalent to telling students. The latter, however, has room for imagination, both in teaching and learning.

The interpretive approach is inspired by the development of new words and language to describe novel scientific discoveries. For example, when Faraday was seeking to describe his pioneering work in electrochemistry, he consulted with his classically trained friend, William Whewell, and invented words such as *electrode, anode, cathode, electrolysis, anion* and *cation* that we still use today. Electrolysis was derived from the Greek *lysis* = loosening and the Latin *electrum* = amber (so called because of the electrostatic charge that could be generated by rubbing amber). Knowledge of the suffix "lysis" has interpretive value to describe the process and can be applied to other chemistry words such as photolysis or catalysis. The prefix "electro" indicates the cause of the loosening, linking to electricity and electrons. Knowledge of the word origin provides context and understanding that there was a reason in human history for the choice of word; it is not just an arbitrary label. Electrolysis has low linguistic demand in the interpretive dimension, if the word roots are known. *Anode* and *cathode* were invented to indicate the flow of current in and out of the electrodes. These words were chosen for somewhat convoluted reasons. Anode originates from the Greek word for sunrise or ascent (*anodos*) and cathode originates from sunset or descent (*kathodos*). Faraday chose these words as he likened the flow of current to the passage of the sun from East to West. As interesting as the origin of these words is, exploring the word origins in this instance is too convoluted and esoteric for many students and would be likely to cause more confusion. Therefore, anode and cathode became labels without interpretive value and have high linguistic demand in the interpretive dimension. Sutton (1992) states that: "*In this way some of the ideas of science get transformed into arbitrary information to be learned; they no longer retain the status of puzzles at all, and scarcely seem to merit being puzzled over. If pupils are exposed to words in that way over and over again, they can get little sense of scientific language as an instrument of interpretation, and little incentive to use it themselves for sorting out ideas*". (p. 51)

An implication is that if students appreciate scientists' linguistic efforts when developing explanations and ideas they may be more confident to participate in personal internal and external negotiations of meaning. In addition, developing decoding skills to interpret new unfamiliar words (Herron, 1996) can make chemical language more accessible. Language is dynamic and evolving and we argue that words such as *anode* and *cathode* (and the related *anions* and *cations*) no longer have interpretive educational value. They do not convey useful meaning but add layers of jargonistic complexity that only serve to make chemistry less accessible to learners.

This dimension also relates to words with specific meanings in science that are different to their everyday meaning. For example, *base*, as defined as a species that is a proton acceptor, scores high in this dimension as this is very different from the everyday meaning such as "the bottom part of something". Activity 9.2 explains how to apply this idea to think about the interpretive value of words used in any curriculum.

Activity 9.2 – Interpretive Words

Exploring the etymological origins of words in chemistry can reveal interpretive value or provide explanation as to why we use the word we do. For example, the word *benzene* can be traced back to the original extraction of benzoic acid from the resin of a tree and the Arabic translation of "incense of Java" (Rees, 2016). Exploring these connections can also provide useful cultural links (see Chapter 4).

i) Think of an area of the curriculum and the words that you commonly use.
ii) Sort these words under two headings – "labels" and "interpretive". Under the labels heading place all words that are used simply as labels for things or processes for which the word offers little explanatory value, *e.g.* cation. Under the interpretive column place all words that provide some interpretive value, *e.g.* electrolysis.
iii) Do you spend time explaining and exploring the parts and whole of the word with the students to show how it relates to its meaning?
iv) Do you know the origins of the words in the "labels" column? Maybe there is hidden interpretive value within some of these words or there is value in explaining where the words originate from so they can be contextualised by the students.

(ii) Sub-microscopic

Chemistry is based on the **sub-microscopic** and particulate nature of matter. Abstract sub-microscopic concepts and their associated vocabulary (*e.g.* atoms, molecules, electrons) present major challenges to students and they can find articulating explanations difficult (Rees *et al.*, 2018b). They are required to accept the model of minute theoretical entities, learn about their nature and then learn to use them to provide explanations of chemical phenomena (Taber 2013). Engagement with the sub-microscopic level requires creative thinking and imagination.

Studies have shown that student understanding of the sub-microscopic level is poor (Nakhleh, 1994; Smith and Metz, 1994, Harrison and Treagust, 2002, Chittleborough and Treagust, 2007) and students have difficulty transferring between the macroscopic and sub-microscopic levels (Gabel, 1998). As demonstrated in the interpretive section, the meaning of *electrolysis* can be derived by interpreting the word morphology and knowledge of word roots. However, to gain a deeper understanding of electrolysis requires the learner to operate within the abstract world of ions, electrons and half equations. Consequently, the linguistic demand of *electrolysis* is very high in the sub-microscopic dimension.

Using the words derived for the curriculum area in Activity 9.2, identify those that operate at the sub-microscopic level. How fluent are your students in their use of these words? Are there pedagogical strategies that can be designed to improve fluency? There may also be words that operate across the macroscopic and sub-microscopic levels and may be used loosely when explaining. For example, referring to the element oxygen operates at the macroscopic level whereas referring specifically to oxygen molecules is the sub-microscopic level. However, there may be times when the macroscopic word (oxygen) may be used when understanding is sought at the sub-microscopic level.

(iii) Similarity

The third dimension is **similarity** between different words. In the interpretive dimension, we have seen how common roots of words can be used to decode new and unfamiliar words. On the other hand, similarity between words can also cause confusion when students are processing and developing their own explanations (Rees *et al.*, 2018b). Consider, for example, *solution, solute, solvent, solvation* and *dissolving.* These words are clearly closely related, indicating the potential to derive common meanings. However, when students try to use these words they can quickly become confused. Students require multiple opportunities to interpret and apply these words to become fluent. Similarly, *electrolysis* may be confused with *electrolyte, electrochemistry* or *electrophile.* For novice chemistry linguists it is easy to use similar sounding words with very different meanings that lead to incorrect explanations. Higher linguistic demand in this dimension corresponds to words that can be confused with other similar sounding words. Activity 9.3 describes a simple activity to explore student usage of similar sounding words.

Activity 9.3 – Similarity

Topics with similar sounding words can cause confusion when trying to develop explanations. This activity focuses on these challenges.

i) Choose a chemical phenomenon that requires explaining using similar sounding words such as separating rock salt by dissolving. Highlight the key vocabulary to be included in the explanation without defining the meaning of the words.
ii) Ask groups of 2–3 students to develop an explanation for this process using the key vocabulary.

iii) What do you observe during the discussion? How appropriately and precisely do the learners use the key vocabulary? Are there instances of confusion where similar words are used in the wrong instance?
iv) What strategies could be used to improve fluency in this area?

(iv) Multiple contexts

The final dimension relates to words that can be used in **multiple contexts** with different meanings. Previous studies (*e.g.* Rees *et al.*, 2018a) have consistently highlighted the difficulties learners have with words such as *salt, base* and *neutral* that have an everyday meaning but a different and more precise meaning in science. *Electrolysis* has low demand in this dimension because the word conveys the same meaning in whichever context it is used. Dual meaning vocabulary can score highly in this dimension because of the use of the same word in different contexts. For example, a *base* can refer to substance that will neutralise an acid, the bottom of an object or a specific place.

Activity 9.4 – Multiple Contexts

Words that have different meanings in different contexts can present difficulty for students. This activity is designed to focus specifically on these words.

(i) Present learners with a range of words that have different meanings in an everyday and a chemistry context (*e.g. base, reduction, weak, polar*) and ask them to provide an example using the word in an everyday and a scientific context.
(ii) Is the word used correctly in a scientific context? Discuss with the students the differences between the meaning in an everyday and scientific context.
(iii) What other ways might you have students distinguish between words used in everyday and in chemical contexts?

9.6.1 Visualising Overall Linguistic Demand

As an aid to developing language consciousness and awareness of overall linguistic demand, the interpretive, sub-microscopic, similarity and multiple contexts dimensions can be represented graphically. The linguistic demand of a word is scored between 1 (low) to 10 (high) based on the scorer's judgement for each of the four dimensions. The scores generated are not intended to be consistent across educators but are specific to the learner context and educational setting. They are subjective and are a mechanism to promote reflection amongst educators on the linguistic demand of the words used. Through discussion, agreement on ratings can be achieved *via*

Table 9.2 The linguistic demand of the word "*electrolysis*".

Dimension	Score	Comment
Interpretive	10	High demand for the novice chemist. However, there is the potential to develop understanding of word morphology.
Sub-microscopic	10	Understanding requires operating at the sub-microscopic level.
Similarity	7	There are many similar words *e.g.* electrolyte, electrostatic, cytolysis.
Multiple contexts	2	Use is in similar contexts.

consensus. For example, *electrolysis* (for a novice chemistry language learner) would score high in the interpretive, sub-microscopic and similarity dimensions but low in the multiple context dimension (Table 9.2).

These scores can be represented graphically as shown in Figure 9.4. The graph highlights the dimensions that cause greatest challenge and the shaded area describes the overall linguistic demand of the word.

This technique can be applied to individual words, whole sentences or passages of text. Consider, for example, this definition of a salt:

> "*any chemical compound formed from the reaction of an acid with a base, with all or part of the hydrogen of the acid replaced by a metal or other cation*" (Lexico, 2019).

To understand the meaning there are a further eight scientific words to comprehend (*chemical, compound, reaction, acid, base, hydrogen, metal* and *cation*). Words such as *metal* may be familiar in everyday language although the context here is confusing and the meaning ambiguous. In an everyday sense, *metal* would be thought of at the macroscopic level as a piece of metal

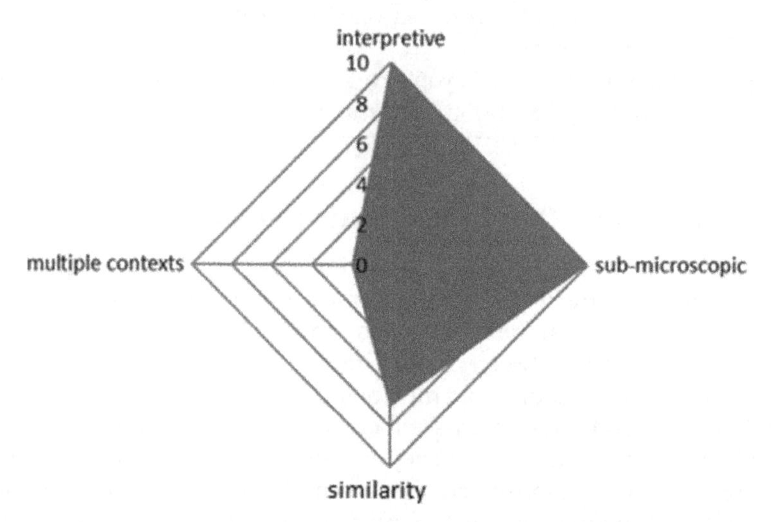

Figure 9.4 Graphical representation of the linguistic demand for "*electrolysis*".

Table 9.3 Linguistic demand of individual words required to define "*a salt*".

	dimension			
Word	interpretive	Sub-microscopic	similarity	Multiple contexts
Chemical	5	1	1	1
Compound	10	1	5	5
Reaction	5	1	7	5
Acid	10	10	1	5
Base	10	10	1	7
Metal (ion)	10	10	1	7
Hydrogen (ion)	10	10	5	1
Cation	10	10	5	1
Total	70	53	26	32

but in this context the definition is referring to metal ions at the sub-microscopic level but this is not specified.

There is similar ambiguity with the reference to hydrogen. What does "with all or part of the hydrogen of the acid replaced" mean? To the novice chemist this could be interpreted as part of the hydrogen itself being replaced. A clearer definition would be "a salt is a chemical compound formed from the reaction of an acid with a base. During the reaction, cations replace hydrogen ions of the acid to form a salt". Each of the words in the definition present challenges to different extents across the four dimensions and can be scored as shown in Table 9.3.

The scores for novice chemistry learners indicate that many of the words are demanding in the interpretive and sub-microscopic dimensions. For example, the word *base* provides no interpretive value to the novice student, is operating at the sub-microscopic level and has different meanings in different contexts. Cumulatively the demand of the entire sentence can be visualised (Figure 9.5).

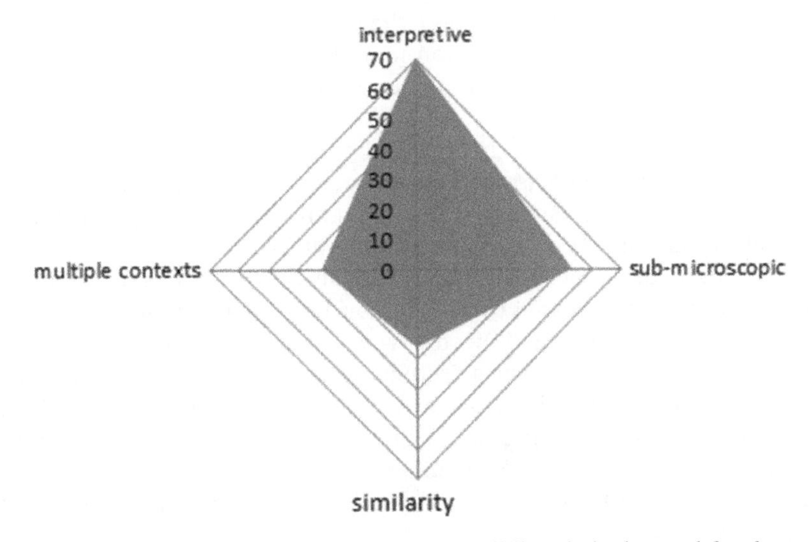

Figure 9.5 Graphical representation of the overall linguistic demand for the words required to define "*a salt*".

The visual representation shows that the interpretive and sub-microscopic dimension are the most significant in this instance. This approach can be taken with any piece of text and highlights the issues for educators so that pedagogy can be targeted to address the challenges.

Activity 9.5 – Reducing Linguistic Demand

This activity is designed to develop understanding with lower linguistic demand.

(i) Choose a definition from a textbook and analyse it for overall linguistic demand based on the four dimensions – represent this graphically.
(ii) Devise a new definition with lower overall linguistic demand and also represent this graphically.
(iii) Try using this new definition in parallel with the formal definition until the latter is established. The definition with lower demand can be removed.

9.6.2 Teaching with Respect to the Dimensions

Laszlo (2013), amongst others, has argued that the language of chemistry should be taught as a language in its own right and science teachers should see themselves as language teachers. However, most chemistry teachers see language and subject content as dichotomous (Markic and Childs, 2016) and do not perceive their role as teaching language, even chemical language. Chemistry teachers are likely to be experts in their field and have mastered the specialised language themselves – the challenges to reach that point may often be forgotten. This may be even more the case at undergraduate level, where lecturers assume that their students already have a grasp of the basic language of chemistry.

Appreciating that words can present linguistic demand in different dimensions requires the creative teacher to equip themselves with a range of appropriate teaching strategies. Table 9.4 summarises a range of practical approaches that have been suggested in the literature.

For words that present high interpretive demand it is important to explore whether this can be reduced by investigating word origins in terms of morphology and how the word came to be used in science. Sutton (1992) recognises that selectivity is important and this cannot be undertaken on all occasions. As we have already seen with words like *anode* and *cathode*, exploring the origins of these words can be confusing and add to cognitive load. Explicit word activities (Activity 9.6) can enable learners to improve their understanding of word families and the links between words.

Table 9.4 Summary of language and literacy focused pedagogical strategies. Reproduced from Rees, Kind & Newton, 2019 with permission from John Wiley and Sons, © 2019 Wiley-VCH Verlag GmbH & Co. KGaA, Weinheim.

Understanding of scientific language		
Strategy	Explanation	Reference
Combining vernacular and non-vernacular (double talk)	Use of two versions of a term. For example, when developing understanding of the word "classified" the teacher states "you've classified, you've decided that things go in different groups".	Brown and Spang (2008) Lemke (1989)
Decoding skills	Developing knowledge of Latin and Greek roots so that new unfamiliar vocabulary can be decoded.	Herron (1996) Rees *et al.*, (2018a)
Multimedia	Engagement in images, animations, virtual models.	Mayer (2014) Rees *et al.*, (2018a)
Word games	Use of word games to increase familiarity with chemical vocabulary.	Herron (1996) Rees *et al.*, (2018a)
Exemplification – corpus linguistics	Use of a scientific text database to illustrate word usage in context.	Rees *et al.*, (2018a)
Cues	Inferring meaning from context.	Herron (1996)
Interpretative meaning	Exploring the stories behind the names for words to enhance their value as interpretive tools and not just labels for processes or substances.	Sutton (1992) Rees *et al.*, (2018a)
Developing initial understanding of scientific ideas		
Strategy	Explanation	Reference
Use of vernacular prior to non-vernacular	Where possible, ideas are introduced using vernacular language prior to introducing non-vernacular language. This can reduce anxiety and ensure that content understanding is established prior to linguistic understanding.	Brown and Spang (2008)
Discursive strategies	Guided and structured activities to promote meaningful discussion incorporating and extending language usage.	Lemke (1989) Edwards and Mercer (2013) Carlsen (2007)
Engaging with scientific text		
Strategy	Explanation	Reference
Close reading strategies	Development of analytical reading strategies to focus reading.	Brown and Concannon (2016)
Scaffolding activities	Staged reading comprehension activities to develop analytical reading.	Wellington and Osborne (2001)
Rules of inference	Developing understanding of information that can be inferred such as from chemical equations.	Herron (1996)

Activity 9.6 – Interpretive Word Tree

Exploring the common roots of words can be a useful way to develop meaning and enable students to "decode" unfamiliar vocabulary. This creative/constructive activity for students is one way of focusing on this with a mind-map based activity.

i) Using a sketch of a tree and its roots reaching through the ground, the root word is placed at the base of the trunk *e.g.* "hydro" with its definition (relating to water).
ii) Students are then challenged to use divergent and associative thinking to come up with as many words as they can that contain this root *e.g. hydrogen, hydrocarbon, hydrophobic.*
iii) These words are written at the end of each of the roots on the diagram with the definition of the word written along the root.

The resulting diagram provides a visual way of summarising and organising a whole word family and helps students see connections across topics.

Note: This can be extended to enable the students to find as many words containing this root as possible on the internet, for example.

For words that have high demand in the sub-microscopic dimension, teaching strategies that encourage creative thinking and visualisation are helpful. This can include role play such as allowing students to "represent" electrons themselves or modelling atomic structures using classroom materials or *via* computer simulations (Activity 9.7).

Activity 9.7 – Representing Electronegativity

In instances where sub-microscopic linguistic demand is high, teaching strategies that help students relate to and visualise sub-microscopic interactions are recommended. This role play activity is used to explore creativity.

Electronegativity is described as a measure of the ability of an atom to attract a bonded pair of electrons to itself. It is affected by the atomic number (*i.e.* nuclear charge) and distance from the nucleus to the shared pair of electrons. Bonds between atoms with differing electronegativity will result in polar covalent or ionic bonds forming.

(i) Ask two students to arrange themselves as a hydrogen atom with one proton and one electron. Then ask 18 students to be a fluorine atom with nine protons and nine electrons.
(ii) Ask them to arrange themselves as a molecule of hydrogen fluoride and then to consider the relative ability of the nuclei to pull the

bonded pair of electrons towards themselves. The physical representation and the ensuing discussion will engage the learners in considering the relative effect of the size of the nucleus (nine people pulling instead of one) and distance from the nucleus to the bonded pair of electrons.

(iii) Students can then be invited to select another molecule for representation in this way.

Meanings of words with high demand in the similarity and multiple contexts dimensions can be taught by facilitating opportunities to promote word usage. These include dialogical approaches (both teacher – learner and learner – learner), word games (as in Activity 9.8), peer feedback, scaffolded tasks and exposure to word usage in multiple contexts (Rees *et al.*, 2013).

Activity 9.8 – Word Games

Where words present high demand in the similarity and multiple contexts dimension, provide multiple opportunities to practise word use. This activity describes how word games can be incorporated to provide students with the opportunity to use chemistry words.

(i) Think of a popular word game that you may have played outside of the classroom. This could be games like drawing a description of a word without saying anything, describing a word without being allowed to say five key words relating to that word (see below in (ii)), or word association where participants have to quickly say a word associated with the previous word.

(ii) Choose one of these games and think about how it can be modified to become a classroom activity. For example, using mini-whiteboards, students write a key word at the top and then five words beneath that are related to the key word (*e.g.* key word = *atom*: five related words could be *proton, neutron, electron, particle, molecule*). The boards are then collected and someone has to try to describe the key word to the rest of the class without saying any of the other five words.

9.7 Conclusion

This chapter has demonstrated the varied challenges that the language of chemistry present and, therefore, students' ability to think creatively about the subject. We have highlighted how chemistry words present linguistic demand in four different dimensions: interpretive, sub-microscopic, similarity and multiple contexts. The challenge of individual words then becomes cumulative with a range of chemistry words needed to explain a new word. In consequence, chemistry educators should develop an awareness of their use of language (language consciousness) and create a range of suitable pedagogical strategies to ensure that vocabulary and the language of chemistry is taught explicitly, rather than by happenstance.

References

Ali M. and Ismail Z., (2006), Comprehension Level of Non-technical terms in Science: Are we ready for Science in English, *J. Pendidik dan Pendidikan*, 73–83.

Ausubel D. P., (1963), Cognitive structure and the facilitation of meaningful verbal learning, *J. Teach. Educ.*, **14**, 217–221.

Beck I., McKeown M. G. and Kucan L., (2002), *Bringing Words to Life: Robust Vocabulary Development*, New York: Guilford.

Brown B. A. and Spang E., (2008), Double talk: Synthesizing everyday and science language in the classroom, *Sci. Educ.*, **92**(4), 708–732.

Brown P. L. and Concannon J. P., (2016), Students' perceptions of vocabulary knowledge and learning in a middle school science classroom, *Int. J. Sci. Educ.*, **38**(3), 391–408.

Byrne M., Johnstone A. H. and Pope A., (1994), Reasoning in science: a language problem revealed?, *School Sci. Rev.*, **75**, 103.

Carlsen W., (2007), Language and Science Learning, in W. Carlsen, S. Abell and N. Lederman (ed.), *Handbook of Research on Science Education*, London: Routledge, pp. 57–74.

Cassels J. and Johnstone A. H., (1980), *Understanding of Non-Technical Words in Science: A Report of a Research Exercise*, Royal Society of Chemistry.

Cassels J. and Johnstone A., (1983), The meaning of words and the teaching of chemistry, *Educ. Chem.*, **20**(1), 10–11.

Cassels J. and Johnstone A. H., (1985), *Words that Matter in Science: A Report of a Research Exercise*, Cambridge: Royal Society of Chemistry.

Childs P. E. and O'Farrell F. J., (2003), Learning science through English: An investigation of the vocabulary skills of native and non-native English speakers in international schools, *Chem. Educ. Res. Pract.*, **4**(3), 233–247.

Chittleborough G. and Treagust D. F., (2007), The modelling ability of non-major chemistry students and their understanding of the sub-microscopic level, *Chem. Educ. Res. Pract.*, **8**(3), 274–292.

Cink R. B. and Song Y., (2016), Appropriating scientific vocabulary in chemistry laboratories: a multiple case study of four community college students with diverse ethno-linguistic backgrounds, *Chem. Educ. Res. Pract.*, **17**(3), 604–617.

Dahl R., (2007), *The BFG*, UK: Penguin.

Edwards D. and Mercer N., (2013), *Common Knowledge (Routledge Revivals): The Development of Understanding in the Classroom*, London: Routledge.

Eiss A. F., (1961), Problems in semantics of importance in science teaching, *Sci. Educ*, **45**(4), 343–347.

Faraday M., (1865), A course of six lectures on the chemical history of a candle; to which is added a lecture on Platinum... delivered during the Christmas Holidays of 1860–1, W. Crookes. (ed.).

Farrell M. P. and Ventura F., (1998), Words and understanding in physics, *Lang. Educ.*, **12**(4), 243–253.

Fredericks S., (2005), Cognitive reflection and decision making, *J. Econ. Perspect.*, **19**(4), 25–42.

Gabel D., (1998), The complexity of chemistry and implications for teaching, in Fraser B. J. and Tobin K. G. (ed.), *International Handbook of Science Education*, Dordrecht, The Netherlands: Kluwer Academic Publishers, pp. 233–248.

Gardner P. L., (1972), *Words in Science*.

Harrison A. G. and Treagust D. F., (2002), The particulate nature of matter: challenges in understanding the submicroscopic world, in Gilbert J. K., De Jong O., Justi R., Treagust D. F. and Van Driel J. H. (ed.), *Chemical Education: Towards Research-based Practice*, Dordrecht: Kluwer Academic Publishers, pp. 213–234.

Herron J. D., (1996), *The Chemistry Classroom*, Washington, DC: American Chemical Society.

Jasien P. T., (2010), You said "neutral" but what do you mean?, *J. Chem. Educ.*, **87**, 33–34.

Johnstone A. H., (1991), Why is science difficult to learn? Things are seldom what they seem, *J. Comput. Assist. Learn.*, 7(2), 75–83.

Johnstone A. H. and Selepeng D., (2001), A language problem revisited, *Chem. Educ.: Res. Pract. Eur.*, **2**(1), 19–29.

Laszlo P., (2013), Towards teaching chemistry as a language, *Sci. Educ.*, **22**(7), 1669–1706.

Lee O., (2005), Science education with English language learners: Synthesis and research agenda, *Rev. Educ. Res.*, **75**(4), 491–530.

Lee O. and Fradd S. H., (1998), Science for all, including students from non-English-language backgrounds, *Educ. Res.*, **27**(4), 12–21.

Lemke J. L., (1989), *Using language in the classroom*, New York: Oxford University Press.

Lexico, (2019), Salt, Available at https://www.lexico.com/en/definition/salt.

Markic S. and Childs P. E., (2016), Language and the teaching and learning of chemistry, *Chem. Educ. Res. Pract.*, **17**(3), 434–438.

Marshall S., Gilmour M. and Lewis D., (1991), Words that matter in science and technology, *Res. Sci. Technol. Educ.*, **9**(1), 5–16.

Mayer R. E., (2014), Incorporating motivation into multimedia learning, *Learning and Instruction*, **29**, 171–173.

Montgomery S. L., (1996), *The Scientific Voice*, New York, NY: Guilford Press.

Nakhleh M. B., (1994), Students' models of matter in the context of acid–base chemistry, *J. Chem. Educ.*, **71**, 495–499.

Nyachwaya J. M., (2016), General chemistry students' conceptual understanding and language fluency: acid – base neutralization and conductometry, *Chem. Educ. Res. Pract.*, **17**(3), 509–522.

OECD, (2015), PISA 2015 - Draft Science Framework. Available from http://www.oecd.org/pisa/pisaproducts/Draft%20PISA%202015%20Science%20Framework%20.pdf.

Oyoo S. O., (2017), Learner outcomes in science in South Africa: role of the nature of learner difficulties with the language for learning and teaching science, *Res. Sci. Educ.*, **47**(4), 783–804.

Pekdag B. and Azizoglu N., (2013), Semantic mistakes and didactic difficulties in teaching the "amount of substance" concept: a useful model, *Chem. Educ.: Res. Pract.*, **14**, 117–129.

Perfetti C. A. and Hart L., (2002), The lexical quality hypothesis, *Precursors of Functional Literacy*, **11**, pp. 67–86.

Pickersgill S. and Lock R., (1991), Student Understanding of Selected Non-Technical Words in Science, *Res. Sci. Technol. Educ.*, **9**(1), 71–79.

Postman N. and Weingartner C., (1971), *Teaching as a Subversive Activity*, Delta.

Rees S. W., (2016), The Stories behind the words, *Education in Chemistry*, Available at: https://edu.rsc.org/feature/the-stories-behind-the-words/2000132.article.

Rees S. W., Bruce M. and Nolan S., (2013), Can I have a word please–Strategies to enhance understanding of subject specific language in chemistry by international and non-traditional students, *New Dir.*, **9**(1), 8–13.

Rees S. W., Kind V. and Newton D., (2018a), Can language focussed activities improve understanding of chemical language in non-traditional students?, *Chem. Educ. Res. Pract.*, **19**(3), 755–766.

Rees S. W., Kind V. and Newton D., (2018b), The Development of Chemical Language Usage by "Non-traditional" Students: the Interlanguage Analogy, *Res. Sci. Educ.*, 1–20.

Rees S. W., Kind V. and Newton D., (2019), Meeting the challenge of chemical language barriers in university level chemistry education, *Isr. J. Chem.*, **59**(6–7), 470–477.

Smith K. J. and Metz P. A., (1994), Evaluating Student Understanding of Solution Chemistry through Microscopic Representations, *J. Chem. Educ.*, **73**(3), 233–235.

Snow C. E., (2010), Academic language and the challenge of reading for learning about science, *Science*, **328**(5977), 450–452.

Song Y. and Carheden S., (2014), Dual meaning vocabulary (DMV) words in learning chemistry, *Chem. Educ. Res. Pract.*, **15**(2), 128–141.

Sutton C. (1992). *Words, Science and Learning*, Open UP.

Taber K. S., (2013), Revisiting the chemistry triplet: drawing upon the nature of chemical knowledge and the psychology of learning to inform chemistry education, *Chem. Educ. Res. Pract.*, **14**(2), 156–168.

Taber K. S., (2015), Exploring the language (s) of chemistry education, *Chem. Educ. Res. Pract.*, **16**(2), 193–197.

Taber K. S. and Coll R. K., (2002), Bonding, in *Chemical education: Towards research-based practice*, Dordrecht: Springer, pp. 213–234.

Tao P. K., (1994), Comprehension of non-technical words in science: The case of students using a 'foreign' language as the medium of instruction, *Res. Sci. Educ.*, **24**(1), 322–330.

Ünsal Z., Jakobson B., Wickman P. O. and Molander B. O., (2018), Gesticulating science: Emergent bilingual students' use of gestures, *J. Res. Sci. Teach.*, **55**(1), 121–144.

Vladušić R., Bucat R. B. and Ožić M., (2016), Understanding ionic bonding–a scan across the Croatian education system, *Chem. Educ. Res. Pract.*, **17**(4), 685–699.

Vygotsky L. S., (1962), *Thought and Language*, Cambridge Mass.: MIT Press.

Wellington J. J. and Osborne J., (2001), *Language and Literacy in Science Education*, Buckingham: Open University Press.

CHAPTER 10

Assessing Creativity

10.1 Assessing, Recognising, or Evaluating Creative Thinking?

Being creative means producing something more or less novel and appropriate – better still if it is also in some way surprising. Some reject the notion of grading someone's attempt at being creative, partly on the grounds that it is about a mode of thought which cannot be accessed. Others point to the unpredictable nature of potential solutions to a problem – what mark scheme can fit the unpredictable (*e.g.* Rogers and Fasciato, 2005)? Still others argue that assessment upsets or suppresses the unconcerned frame of mind from which creative thinking benefits (*e.g.* Burnard and White, 2008). Perhaps we need to move away from an assessment which attempts to say that one creative idea is, say, 2% better than another. Instead, we may need to focus on *recognising* creative thinking when we see it and *evaluating* its qualities in terms of strengths and weaknesses which inform formative feedback and indicate possible avenues for development. There may be times when a teacher is called on to provide a summative assessment, and there are strategies which can help that process, but fine scale precision may be difficult to achieve (or believe) (Newton, 2012). With these cautions in mind, we offer some thoughts on recognising and evaluating creative teaching and students' creative thinking in chemistry.

10.2 Evaluating Creative Teaching in Chemistry

Creative teaching is an example of what has been called professional creativity, drawing on specialist skills and knowledge which underpin teaching (here, the teaching of chemistry) (Kaufman and Beghetto, 2009). These include formal knowledge (*e.g.* knowledge of chemistry), pedagogical knowledge

Advances in Chemistry Education Series No. 4
Creative Chemists: Strategies for Teaching and Learning
By Simon Rees and Douglas Newton

Published by the Royal Society of Chemistry, www.rsc.org

(*e.g.* teaching in general and, specifically, teaching chemistry), and subject specific matters (*e.g.* programmes of study, rules and regulations for safe practices). In teaching, however, these are also supported to a large degree by informal knowledge and skills associated with working with people and, in particular, with students. These are brought together to solve a practical teaching problem or achieve a goal which could range in magnitude from something small and local (micro) to something large and all-embracing (macro). Table 10.1 illustrates this with the goal of enhancing the *understanding* of a particularly difficult topic, and also for the goal of *engaging* students in learning (Newton, in press). For example, at the micro level, with the goal of supporting understanding, a teacher might construct an analogy to clarify a chemical concept such as collision theory, visualise molecular structures, or explore the origins of a key word to make its meaning clearer and more memorable for the students. At an intermediate level, with the goal of enhancing student engagement, a teacher might humanise the topic by setting it in the context of its discovery (*e.g.* see Section 5.1). Practising teachers most often have opportunities for creative work in the micro and intermediate levels of this model, but occasionally, their ideas find favour more broadly and spread into other curriculum areas. In the 19th century, Armstrong's heuristic approach to chemistry teaching brought the value of 'the scientific method' to the fore, although it was slow to find fertile ground until the 20th century with projects like Nuffield Science.

Table 10.1 Two examples of goals (represented by fostering understanding and fostering cognitive engagement in learning), offering opportunities for creative teaching at different levels of generality, micro, intermediate, and macro. An episode is a self-contained teaching unit (*e.g.* a lesson) intended to support learning.

Underpinning knowledge, skills and experience	**Goal**	**Micro:** *Intra-episode devices*	**Intermediate:** *Episode or episode sequence*	**Macro:** *Cross-curricular concepts*
Informal *e.g.* through interaction; student life; various media; siblings	**Understanding**	*e.g.* Aide memoire; analogy; etymology of terms.	*e.g.* Flipped teaching; blended learning; MOOCs; problem-based learning.	*e.g.* Inquiry-based learning; communities of thinking.
via training; guided practice; teacher observation **Formal**	**Engagement**	*e.g.* An unusual artefact/ puzzle; activity for developing self-efficacy.	*e.g.* Need satisfaction; humanised content or approach; authentic learning.	*e.g.* Sustainability education; values education.

Learning can benefit from innovation anywhere along the continuum from micro to macro. A teacher who wants to be more effective should clarify the goal, identify the specific need, and then play with ideas to satisfy that need (Table 10.1). Talking with a colleague may also help to produce a potential solution. Evaluation of the product is straightforward: the tentative solution is tried, and, if there is the desired learning, the innovation has been successful. It may not, however, be a matter of success or failure. It may be that the idea works well with some of the class but not all, or it may work moderately well with all, but fall short of the desired mark. Further reflection may identify the likely cause and adjust the idea to improve its effect. In short, the process can be an iterative one (paralleling that commonly used in technological design). Here, a teacher evaluates the innovation him or herself – an appropriate course of action where a teacher is responsible and accountable for the quality of learning which takes place under his or her jurisdiction. Teachers might ask themselves: Was it imaginative? Was it appropriate or fit for purpose? Did it achieve the goal? Was it also a pleasing or rewarding lesson? If the session did not achieve its ends, what might have been different?

There are occasions when a teacher's effectiveness may be appraised by someone else, often a senior colleague, a mentor, or a school inspector. Such an appraisal is generally wider than an interest in innovation or creative teaching, but it may identify opportunities where students could benefit from these. A mentor, for instance, may coach a novice by example, taking the novice through the need, tentative ideas, idea development, and idea application. Both novice and mentor may evaluate the success of the idea and compare notes. Reflection on the process as well as on the product of creative thought may also prove useful. If so, some of what follows about fostering students' creative thinking may also be relevant.

10.3 The Formative Evaluation of Chemistry Students' Creative Thinking

Providing opportunities in imaginative ways for students to practise thinking creatively in chemistry is only the first step. Going beyond that, we might ask: How effective are those opportunities? Is the students' competence increasing? Which students are not making progress? Why are they not making progress? What might be done to help them develop their competence? To answer these questions, there needs to be some form of evaluation or assessment of the students' thinking. Not everyone feels comfortable with this when it means giving some numerical score for a piece of 'creative' work. They argue that creative ideas are not like facts to be accumulated and counted, there may be no right answers, and telling students they have 'failed' the test may make them give up, drop out, or live only in the world of information gleaned from other sources (*e.g.* de la Harpe and Peterson, 2008). These are good reasons for being sceptical about the precision of percentages, and for trying to foster a belief that effort and practice can make a difference, but it helps if a teacher knows *where* that effort is needed and *what* that practice should be. Clues to inform the process of teaching can only come from reflection on the products and processes of students' thinking (Newton, 2012).

10.3.1 Clues from Observations of Products and Processes

In teaching, the product is the plan of action that the teacher constructs with the aim of achieving certain learning goals. A teacher might reflect on that plan and feel more or less pleased with some aspects and dissatisfied with others. Reflection on the reasons for the feelings of dissatisfaction may lead to improvements. The process is the collection of mental and physical acts that constructed or created the plan. Again, a teacher may reflect on these, and ask: Is there a better way of planning? In the same way, the product of a student's thinking in chemistry may produce:

- a Why? question for the problem space;
- a tentative explanation from which a prediction is made in the hypothesis space;
- a design for a test of that prediction in the experimental space;
- a solution to a practical problem in the application space.

Scrutiny of these products may locate strengths and weaknesses in creative thinking. How a student constructed these products, is not, of course, open to direct scrutiny, but a teacher may find clues to the creative process by observing actions and behaviours. For instance: Was the student clear about the goal before proceeding? Did the student seek ready-made answers, or was there an attempt to construct them independently? Did the student dismiss ideas early, or explore them for their potential? Was there a tendency to persist with an idea unreasonably? The aim is to find impediments to students' creative thinking so that support and strategies may be provided to develop their competences. In addition, we would probably also want to see a disposition or an unprompted tendency to think in this way. This may develop with practice, but it may also come from an interest in the topic, which takes us back to the need for imaginative teaching that generates such interest. Table 10.2 illustrates some behavioural clues which can inform a formative evaluation and guide advice. Petty has described six phases in the creative process which could also point to such behaviours: Inspiration (ideation), Clarification (goal focusing), Distillation (idea selection), Incubation (unconscious processing over time), Perspiration (effort and persistence), and Evaluation (review and reflection) (see Moseley *et al.*, 2005). In practice, the process is rarely orderly, and in the classroom, time for incubation may be short.

From such observations, a profile may be compiled in the form of a simple 'radar' chart with as many axes as attributes of interest (*e.g.* Figure 10.1). This has the advantage of revealing, rather than concealing by combination,

Table 10.2 Examples of creative thinking activities and behaviours that may need support.

Students working with	Behaviours
A question to answer/ a problem to solve	Clarifying, analysing, identifying goals and criteria for success.
Knowledge exploration and development	Identifying what is relevant and known, what needs to be known, how it will become known.
Tentative idea construction	Adopting a non-analytical frame of mind; drawing on relevant experience, being open-minded about ideas; suspending judgment about idea suitability; being open to alternatives; using pencil-and-paper to record and/or play with ideas.
Idea development	Choosing a promising idea; manipulating an idea to satisfy the goal; being prepared to consider another idea; testing ideas or their parts; considering practicality and feasibility; checking for unexplored aspects.
The answer or solution	Working for completeness; checking against the success criteria (appropriateness); presenting the answer/ solution with clarity; recognising that entire goal satisfaction is not always possible.
Dispositions	Showing: self-motivation, self-reliance, a reasonable persistence, flexible working and thinking, collaborative skills if needed, avoidance of undue haste, being prepared to go back and rethink.

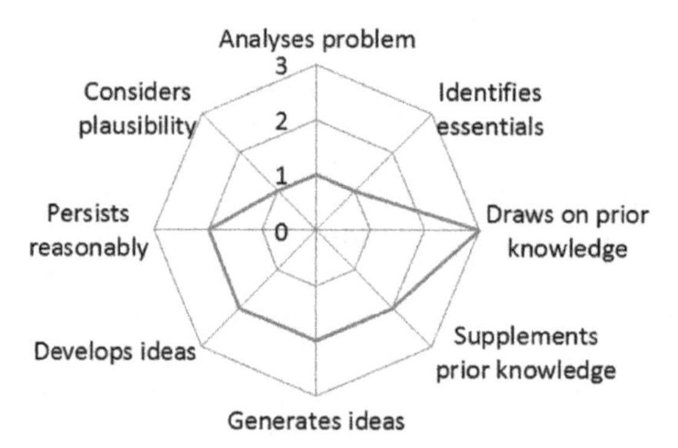

Figure 10.1 An illustrative chart showing a student's record. The level of evidence ranges across: None seen 0, Sometimes 1, Often 2, Always 3. Different behaviours (the spokes or axes) may be seen as appropriate (*e.g.* 'Finds a problem to solve' may suit some expectations), and there may be more or less of them).

strong kinds of thinking and those kinds of thought in need of development and practice. Some students may also find it useful to use this to support their self-evaluation of their creative thinking.

10.4 The Summative Evaluation of Chemistry Students' Creative Thinking

10.4.1 Teacher Judgment

Evaluating creative competence in subjects like chemistry should not be confused with assessing general creative capabilities. The latter has a long history in psychology and may focus on, for instance, the number and variety of ideas, and then use these as indicators of someone's broad creative ability. But, generally being able to produce many and diverse ideas does not mean that someone will be universally creative (or creative in chemistry). It has to be recognised that people often find they become or can be creative more readily in one domain than in another. Equally, some point out that it is not many and diverse ideas that matter, just one good idea. The knowledge, processes and skills associated with different domains, differ, sometimes slightly and sometimes widely. When evaluating creative competence in classrooms, it is important to see it as subject specific, albeit the competences may overlap those of other subjects. One student may be creative in art but not in chemistry; another may be creative in chemistry but not in mathematics, but a student could be creative in, for instance, organic and inorganic chemistry.

A summative account of a student's strengths may be compiled from observation of behaviours, like those in Table 10.2, but care is needed not to base this on only one or two tasks. Creative thinking sometimes involves risk taking: a student may not always play safe, and, instead, tries what on the face of it looks like a novel, simple, clever, pleasing, and economical solution to a problem. But, in its realisation, something which could not have been reasonably anticipated by the student concerned may make it fail. An assessment averaged over several tasks, therefore, is more likely to reflect the student's creative competence.

One way of 'weighing' students' creative ideas (the products) which finds a lot of favour is to rank them from what seems most to least creative. This has been found to work well when done by teachers familiar with students' capabilities and who are also familiar with the subject. Some time ago, Amabile (1996) demonstrated that when such teachers rank products according to their intuitive judgements of creativity, there is generally a very high level of agreement. It seems that teachers in tune with students and their subject can recognise imaginative and appropriate thinking when they see it.

Putting students' ideas in rank order, however, becomes increasingly difficult as the number of students increases. Kokotsaki and Newton (2010) tested a diamond ranking approach to reduce the labour and time. This assumes that only a small number of students will show a very high (or very low) level of creativity, while more will tend to be moderately creative (and uncreative). On this basis, students' work is sorted into a diamond pattern, as in the example in Figure 10.2. Ranking students identifies the strong and the weak (on a given occasion) but it does not, of course, allocate grades or marks. Over time, a

A
B B B
C C C C
D D D D D
E E E E
F F F
G

Figure 10.2 A diamond ranking of 21 products, *A* being considered to be most creative, *D* being seen as of middling creativity, and *G* being least creative.

teacher may feel confident that Student A is exceptional and deserves a mark between 90 and 100%, while Student G is much less competent, but not as weak as others in the past so deserves a mark between, say, 20 and 30% (an intuitive process of norm referencing). Again, some averaging over several products is needed as one alone may not be a reliable indicator of a particular student's competence. At the same time, it has to be remembered that a student's creative thinking may be novel and plausible but is unlikely to match that of the practising chemist, or will always produce solutions like those generally current amongst chemists. This is an occasion where being 'wrong' need not be a fault – imaginative, believable ideas deserve credit (just as they have in the history of science until they were displaced by more believable ideas). There can be a temptation to see only the 'right' explanation as creative. Knowledge of students working at the level of thinking under scrutiny, and knowledge of the subject are essential. For this reason, novice teachers may find it helpful to work with an experienced teacher (Newton, 2010). Others are likely to find it helpful to work collaboratively on at least a sample of students' work, giving particular attention to those where they disagree on their position in the diamond. Finally, for a particular group of products, a teacher may feel that the pattern should not be a symmetrical: 1, 3, 4, 5, 4, 3, 1 formation, like that of Figure 10.2. It could be, for instance: 1, 4, 7, 5, 3, 1 for a more competent group of students).

10.4.2 Seeking Objectivity

Other approaches to collecting evidence that have been suggested include the gradual compilation of a portfolio which shows the generation of ideas (ideation), the justified development of ones that are promising, and their presentation as credible solutions (Allen and Coleman, 2011). This approach would lend itself to Clark's use of research projects at the undergraduate level (which could be readily adapted using mini-projects or tasks at other levels of education) (Clark, 2015).

Some could argue that for creative thinking to be valued, it must be included amongst kinds of thinking and learning that are already assessed by some form of test set by an external agency. The student's performance may then be graded according to a pre-determined mark scheme, and the grade used as an indicator of creative competence. (For that reason, Blamires and Peterson (2014, p. 156) include assessment in a list of 'enablers' of creativity, 'those circumstances and support mechanisms that make creativity more likely to thrive'. Other enablers include culture, curriculum, individual skills, teaching and learning approaches, teacher attitudes, resources and tools.) Barbot *et al.*, (2011) describe some of the challenges associated with such assessments, and describe the development of a test of creativity in particular subject domains, but these are not tied to a particular programme of study. Most teachers will find themselves working with a given programme and the test will relate to that.

Such tests could have various forms. For example, students might be given a problem and asked to generate a plausible, tentative hypothesis, followed by a practical way of testing it. Given that one exercise may not give the student an opportunity to demonstrate a wide range of competences, the task may include several such problems, but this would take more time. As a project, the problem may be set in advance, and the student would then work on it over several weeks (although some external agencies may feel uncertain about the authenticity of a student's submission when it has not been completed under examination conditions). Whatever the form of the test, submissions have to be graded in some way. Although examiners could use the consensus approach outlined above, with moderation to align the levels, they may be given some form of mark scheme. This mark scheme should recognise the essential unpredictability of creative thinking, so it could not be of the form that gives a certain number of marks for a particular answer. There would always be an answer that was plausible but not anticipated. This means that criteria would have to be general. For example, take: 'an imaginative solution to the problem'. A five-point assessment scale might be:

- No solution: 0;
- A reproductive solution: 1;
- Connects cognitively near concepts: 2;
- Connects with a cognitively remote concept: 3;
- Forms a coherent structure connecting near concepts and at least one remote concept: 4.

Such a scale seems to offer some objectivity but could be slow to use as examiners would need to analyse each solution quite carefully. This might be made easier if examiner guidance included vignettes of students work to illustrate each level. Those who are willing to accept a degree of intuitive or unconscious judgement of creative qualities may begin with a diamond sort of a random sample of students' work and reflect on the attributes which they feel made *A* the best, and *C* better than *D*, *etc.* These attributes would then be listed, described, and assigned levels of achievement to use in grading all the students' work. (Again, a set of vignettes might be helpful.) Occasionally, as grading progresses, a level of achievement may need adjustment to accommodate a different way of looking at the problem.

10.5 Conclusion

The formal assessment of creative thinking in specific disciplines like chemistry does not seem to be well developed yet. Nevertheless, the consensus approach, sometimes described as the 'gold standard' for assessing creativity in such domains (Baer, 2014), does offer a practical approach which may satisfy society's need for examination grades, provided that the limitations of grading, particularly that based on one task, are kept in mind. At the same time, it must always be kept in mind that a single grade can conceal more than it reveals.

References

Allen B. and Coleman K., (2011), The creative graduate: cultivating and assessing creativity with eportfolios, *Proceedings Ascilite 2011*, Hobart, pp. 59–69. Available at: http://www.ascilite.org.au/conferences/hobart11/procs/Allen-full.pdf.

Amabile T. M., (1996), *Creativity in Context*, Boulder: Westview Press.

Baer J., (2014), The gold standard for assessing creativity, *Int. J. Qual. Assur. Eng. Technol. Educ.*, **3**(1), 81–93.

Barbot B., Besançon M. and Lubart T. I., (2011), Assessing Creativity in the Classroom, *Open Educ. J.*, **4**(1: M5), 58–66.

Blamires M. and Peterson A., (2014), Can creativity be assessed?, *Cambridge J. Educ.*, **44**(2), 147–162.

Burnard P. and White J., (2008), Creativity and performativity, *Br. Educ. Res. J.*, **34**(5), 667–682.

Clark T. M., (2015), Fostering Creativity in Undergraduate Chemistry Courses with In-class Research Projects, in Charyton C. (ed.), *Creativity and Innovation Among Science and Art*, London: Springer.

de la Harpe B. and Peterson F., (2008), A model for holistic studio assessment in the creative disciplines, in *ATN Assessment Conference 2008*. Available at: https://ojs.unisa.edu.au/index.php/atna/article/view/339.

Kaufman J. C. and Beghetto R. A., (2009), Beyond Big and Little: The Four C model of creativity, *Rev. Gen. Psychol.*, **13**(1), 1–12.

Kokotsaki D. and Newton D. P., (2010), Recognizing creativity in the music classroom, *Music Educ.*, 3(4), 91–508.

Moseley D., Baumfield V., Elliott J., Gregson M., Higgins S., Miller J. and Newton D. P., (2005), *Frameworks for Thinking*, Cambridge: Cambridge University Press.

Newton D. P., (2010), Assessing the creativity of scientific explanations in elementary science: an insider–outsider view of intuitive assessment in the hypothesis space, *Res. Sci. Technol. Educ.*, **28**(3), 187–201.

Newton D. P., (2012), Recognizing creativity, in Newton L. D. (ed.), *Creativity for a New Curriculum*, London: Routledge, pp. 108–118.

Newton D. P., (in press). The creative teacher in a changing world, in Harpaz Y. (ed.), *Teaching in the 21st Century* (working title), Jerusalem.

Rogers M. and Fasciato M., (2005), Can creativity be assessed?, presented at the BERA Annual Conference, University of Glamorgan, 14–17 Sept. 2005. Available at: leeds.ac.uk/educol/documents/150029.htm.

CHAPTER 11

Why Creativity Matters

We have pointed out that robots in one form or another are not unusual in the workplace and, with advances in artificial intelligence, their use is expanding into everyday life. Wherever actions can be rendered into routines, robots may be constructed to carry them out. Extrapolating this into the future, more and more repetitive occupations will become the province of robots. Only those associated with creative thought have a certain future, a niche that (at least at present) is the preserve of people. This alone gives urgency to the fostering of imaginative thought across the disciplines, and that includes chemistry. The reward will be its contribution to a country's economy, and to the solving of local and global scientific problems, as with the chemical aspects of climate change and pollution. In addition, an increased use of digital teaching aids and teaching robots means that, like all teachers, the chemistry teacher will have to look to their role and what it is that they do that a robot could not.

While such benefits are very important, there is a tendency to overlook something equally important at a more personal level. Some creative competence and a disposition to use it can help to make people autonomous beings. Life can be a succession of problems but, on meeting them, a creative mind is not a helpless mind. Through imagination and the bringing together of ideas, it can construct alternatives and effective solutions (Newton, 2012a, b). In effect, creative competence bestows a resourcefulness and personal efficacy that can help people to adapt in a rapidly changing world. Having such a mental tool is also potentially empowering, reassuring, and supportive of mental health and well-being (Newton, 2016). In addition, as any chemical engineer, biochemist, or forensic scientist will tell you, its successful application can be enormously pleasing, satisfying, and generally emotionally rewarding.

But there is a reason why creativity is important for both society *and* for the individual. Artificial intelligence and human intelligence are not the

Advances in Chemistry Education Series No. 4
Creative Chemists: Strategies for Teaching and Learning
By Simon Rees and Douglas Newton

Published by the Royal Society of Chemistry, www.rsc.org

same. Artificial intelligence is tireless, constantly observant, and accurate to however many decimal places you want. On the other hand, human intelligence is partly emotional, prone to error, and, at times, erratic. In an age where digital technology is powerful and ubiquitous, there is a danger that human intelligence is seen as weak or substandard, and what becomes important in education is *only* what artificial intelligence can do (Newton and Newton, 2019). This could only stifle and limit the possibilities of human intelligence. Fostering creative thought could be an antidote to such a tendency, and, in the process, develop a competence of long term value.

In spite of these societal and personal benefits, creativity can have a down side (Cropley *et al.*, 2010). Perpetual innovation threatens society's stability and cohesion, and can even make people anxious and uneasy in an ever-changing world (McLaren, 1999). It has been argued that, while, at times creative activity is worthwhile, it has achieved the unwarranted status of an unqualified virtue, and its pursuit has attracted romantic connotations of heroism (Osborne, 2003). There are even occasions when creativity is malevolent, and the link between creativity and integrity is negative (Beaussart *et al.*, 2013). For instance, some might argue that adding to the range of explosive fireworks lacks moral justification. An overly narrow view of creative activity could even add to our problems, as when it encourages single-use or non-biodegradable plastics, which are bad for the environment, and wastes resources. It could be argued that creativity needs to be directed more towards a make-do-and-mend society (Craft, 2008). And, of course, creative competence can be misdirected, as in the imaginative activity of drug smugglers and coin forgers. Creative competence alone is not enough: it needs to be guided and tempered by wise thinking, moral principles, and a concern for consequences. On this basis, we should say, perhaps, that it is a *defensible* creativity that matters and the mental competences that both support *and* control it should be fostered in education.

The nature of teaching is not fixed. At different times, what it means to be a teacher has been different. For some teachers today, the needs of students and the changing nature of teaching may call for adjustments to what they do, and to how they see their role, and even to their professional identity.

11.1 The Chemistry Teacher's Role

What does it mean to be a chemistry teacher? A teacher's professional identity is complex and varies from teacher to teacher, but what teachers value and see as worthwhile in their role is often shared by others. For instance, many teachers want to teach because they want to have a positive influence on students' lives, or they see teaching as being about helping students develop skills for adulthood, or it is to pass on the passion they feel for their subjects. Or it may be to put their interest in learning into practice by creating effective learning environments (Olsen, 2008; Mansfield and Beltman, 2014). What it means to be a teacher can contribute significantly to what teachers do in the classroom: in other words, their identity is reflected in their practice (although it may be subject to the constraints of the prevailing school ethos and to societal expectations) (*e.g.* Chang, 2019). Success in such roles can bring with it personal rewards, like job satisfaction, enjoyment, and feeling positive about teaching. It also offers pedagogical rewards, such as enhanced student learning and attainment. Nevertheless, at a more specific level, a teacher could also believe that learning is largely about transmitting information to students who must then reproduce it in an examination. Another teacher may be more concerned about fostering constructive thought. Both may see these as influencing and equipping students for the future, but do they equip their students equally well?

The world is changing and seeing teaching with blinkered vision is not enough. We believe there is a growing need to put a premium on constructing understandings and thinking creatively in both teaching and learning. Teachers themselves can expect to work alongside robots in the classroom, and their roles will change from the lone provider of learning experiences to being a joint teacher and a creative manager of educational provision (Newton and Newton, 2019). Their students will need to be prepared for life in a digital world where routine processes are automated, leaving them to do what AI does less well, namely, think creatively (SCAI, 2018). Preparing students for this kind of world has the potential to contribute positively to their lives now and in the future, it will call for the development of mental skills for use in that future, and it will allow teachers to pass on their passion for their subject through the opportunities they provide for exploration and the construction of meaning. At the same time, the teachers themselves will need to construct new learning environments which are likely to achieve these goals, and, in the process, experience the enormous satisfaction and positive feeling which comes from successful problem solving and creative thinking.

Teachers' professional identities can support creative teaching and learning, although these dimensions may need to be made explicit and open to reflection. We have tried to do this for these in the context of chemistry for both teaching and for learning. We see these two dimensions as being crucial to science education in all its stages if such an education is to meet the needs of a changing world, and those who will have to live and work in it.

We have illustrated the first dimension, that of the teacher, by demonstrating throughout this book, creative approaches to chemistry teaching. This has included: multisensory approaches that enable students to experience chemistry with all their senses; chemistry in a variety of different cultural contexts that promotes new remote connections, challenging accepted chemistry representations such as the Periodic Table; using storytelling and performance to promote curiosity and engagement, practical strategies that develop new approaches, and strategies to help students develop understanding of the language of chemistry. As experienced teachers ourselves, however, we appreciate that the time poor and over stretched teacher may struggle to find the space and capacity to develop new creative ideas and teaching skills. However, in our experience, engagement in creative teaching leads to benefits for the individual teacher, increased job satisfaction and professional recognition.

We have illustrated the second dimension, that of the student, by demonstrating how learning chemistry is so much more than stereotypical notions of complex and confusing experiments described with unfamiliar symbols and formulae. We have demonstrated how chemistry learning can be situated within all aspects of our lives such as: the food we eat, the activities we do, the places where we live and the stories we enjoy and tell. We have shown the creative approaches to learning that can be applied and while, for many, the focus may be on passing the next formal examination, there is intrinsic value in thinking creatively to explore and understand the physical world.

When talking of creativity in science, we sometimes find students who suspect that there is nothing much left to discover. At the same time, a teacher feels unable to find a new way to teach a topic. But, being creative is not really about 'discovery' or 'finding' something: it is a way of thinking, problem solving, and constructing different ways of seeing the world (Newton, 2012b). Erwin Schrödinger expressed it well when he wrote, 'The task is not to see what has never been seen before, but to think what has never been thought before about what you see every day'. This applies equally to the student's learning and the teacher's chemistry lesson.

References

Beaussart M. L., Andrews C. J. and Kaufman J. C., (2013), Creative liars: The relationship between creativity and integrity, *Think. Skills and Creativity*, **9**, 129–134.

Chang K. C.-C., (2019), Examining teacher identity development, *Compil. Transl. Rev.*, **12**(2), 127–172.

Craft A., (2008), Nurturing creativity, wisdom, and trusteeship in education, in Craft A., Gardner H. and Claxton G. (ed.), *Creativity, Widsom and Trusteeship*, Thousand Oaks: Corwin, pp. 1–15.

Cropley D. H., Kaufman A. J., Kaufman J. C. and Runco M. A., (2010), *The Dark Side of Creativity*, Cambridge: Cambridge University Press.

Mansfield C. F. and Beltman S., (2014), Teacher motivation from a goal content perspective, *International Journal of Educational Research*, **65**, 54–64.

McLaren R. B., (1999), Dark side of creativity, in Runco M. A. and Pritzker S. R., *Encyclopedia of Creativity*, San Diego: Academic Press, pp. 483–492.

Newton D. P., (2016), *In Two Minds: The Interaction of Moods, Emotions and Purposeful Thought in Formal Education*, Ulm-Germany: The International Centre for Innovation in Education.

Newton D., (2012a), *Teaching for Understanding*, London: Routledge.

Newton D., (2012b), Creativity and problem solving: an overview, in Newton L. (ed.), *Creativity for a New Curriculum*, London: Routledge, pp. 7–18.

Newton D. P. and Newton L. D., (2019), Humanoid robots as teachers and a proposed Code of Practice, *Front. Educ.*, , DOI: 10.3389/feduc.2019.00125.

Olsen B., (2008), How reasons for entry into the profession illuminate teacher identity development, *Teach. Educ. Q.*, **35**(3), 23–40.

Osborne T., (2003), Against creativity: a philistine rant, *Econ. Soc.*, **32**(4), 507–525.

SCAI (Select Committee on Artificial Intelligence), (2018), *AI in the UK: Ready, Willing and Able?* HL Paper 100, London: HMSO, Available online at: https://publications.parliament.uk/pa/ld201719/ldselect/ldai/100/100.pdf.

Subject Index